FIELD-PROGRAMMABLE GATE ARRAY TECHNOLOGY

FIELD-PROGRAMMABLE GATE ARRAY TECHNOLOGY

edited by

Stephen M. Trimberger
Xilinx

with contributions by

Stephen M. Trimberger
Xilinx

Dennis McCarty
Telle Whitney
Actel

and
The Technical Staff of Altera Corporation
edited by
Robert Hartmann

Springer

Library of Congress Cataloging-in-Publication Data

Field-programmable gate array technology / edited by Stephen M.
Trimberger.
p. cm.
Includes bibliographical references and index.
ISBN: 0-7923-9419-4
Gate array circuits. 2. Programmable logic devices.
Programmable array logic. I. Trimberger, Stephen, 1955 -
TK7895.G36F54 1994
621.39'5—dc20 93-39703
CIP

Trimberger: Field-Programmable Gate Array Technology

First Indian Reprint 2007
Sixth Indian Reprint 2013
Seventh Indian Reprint 2014
Eighth Indian Reprint 2015

ISBN 978-81-8128-603-1

This edition is published by Springer (India) Private Limited, (a part of Springer Science+Business Media), Registered Office: 7th Floor, Vijaya Building, 17 Barakhamba Road, New Delhi – 110 001, India

Printed in India by Rajkamal Electric Press, Plot No. 2, Phase-IV, HSIDC, Kundli, Haryana.

to ross

who had a vision

Contents

Preface xi

Chapter 1. Introduction 1

1.1. Logic Implementation Options 1

1.2. What is an FPGA? 2

1.3. Advantages of FPGAs 4
- Low Tooling Costs 4
- Rapid Turnaround 4
- Low Risk 5
- Effective Design Verification 6
- Low Testing Costs 6
- Standard-Product Advantages 7
- Life Cycle Advantages 8

1.4. Disadvantages of FPGAs 8
- Chip Size and Cost 8
- Speed of Circuitry 9
- Design Methodology 9

1.5. Technology Trends 10
- Density 10
- Speed 10
- Architecture 11

1.6. Designing for FPGAs 11
- Design Migration 11

1.7. Outline of Subsequent Chapters 12
- 1. Introduction 12
- 2. Programming Technology 13
- 3. Device Architecture 13
- 4. Software 13
- 5. The Future 13
- 6. Design Applications 13
- 7. Acknowledgments 13
- 8. References 13

1.8. References 13

Chapter 2. SRAM Programmable FPGAs **15**

2.1. Introduction **15**

2.2. Programming Technology **15**
SRAM Programming 15
Advantages and Disadvantages of SRAM Programming 17

2.3. Device Architecture **19**
Simple SRAM-Programmable FPGA Architecture 19
Design Trade-offs 23
The Xilinx XC2000 Architecture 29
The Xilinx XC3000 Architecture 35
The Xilinx XC4000 Architecture 43
Programming the FPGA 52

2.4. Software **53**
Automated Design Implementation 54
Technology-Specific Synthesis 63
Manual Design 63

2.5. The Future **65**
Programming Technology 65
Architecture 66
Software 66
Partitioning in Space and Time 67
Design Methodology 67

2.6. Design Applications **68**
General Design Issues 68
Counter Examples 70
Efficient Multiplication by a Constant in an Artificial Neural Network 75
Distributed Arithmetic for Signal Processing 77
Applications of Reprogramming 79
A Fast Video Controller 83
A Position Tracker For a Robot Manipulator 84
A Fast DMA Controller 85
Custom Computing Applications 87

2.7. Acknowledgments **90**

2.8. References **91**

Chapter 3. Antifuse Programmed FPGAs 97

3.1 Introduction 97

3.2 Programming Technology 99

3.3 Device Architecture 103
Principles of Programmable Routing 103
Routing Architecture of the Actel FPGAs 108
Act1 Architecture 110
Act2 Architecture 113
Act3 Architecture 117
Programming and Testing 118
Capacity 124
Performance 127

3.4 Software 128

3.5 The Future 132

3.6 Design Applications 133
Designing with ACT1 and ACT2 FPGAs 133
Designing with ACT FPGAs: A TTL Perspective 137
Migrating PLD Designs to FPGAs 140
Synthesis Design Flow 143
Designing Counters with ACT Devices 144
Designing Adders and Accumulators with the ACT Architecture 153
State Machine Design 160
Using FPGAs for Digital PLLs 164
Customer Design Examples 167

3.7 Acknowledgments 168

3.8 References 168

Chapter 4. Erasable Programmable Logic Devices 171

4.1. Introduction 171

4.2. Programming Technology 173
Logic Structures Using EPROM Transistors 175

4.3. Device Architecture **179**
Basic Concepts 179
Macrocell Architecture 180
Logic Array 180
Programmable Flip-Flops 181
Programmable Clock 182
I/O Control Block 182
Design Security 182
Functional Testing 183
Operating Requirements for EPLDs 183
Architectural Evolution in Array-Based PLDs 184
4.3.1 - The "Classic" Family of PLDs 184
Functional Description of the EP1810 184
4.3.2 - The MAX (Multiple Array matriX) Product Family 187
4.3.3 - MAX 7000 195
4.3.4 - MPLDs: Mask-Programmed Logic Devices 200

4.4. Software **204**

4.5. The Future **218**

4.6. Design Applications **224**
4.6.1 MAX 5000 Timing 224
4.6.2 Using Expanders to Build Registered Logic in MAX EPLDs 228
4.6.3 Simulating Internal Buses in General-Purpose EPLDs 233
4.6.4 Fast Bus Controllers with the EPM5016 238
4.6.5 Micro Channel Bus Master and SDP Logic with the EPM5032 EPLD 240
4.6.6 FIFO Controller Using an EPM7096 243
4.6.7 Integrating an Intelligent I/O Subsystem with a Single EPM5130 EPLD 246
4.6.8 Controlling Complex CCD Imaging Systems with the EPS464 EPLD 247

4.7. References **250**

Index **253**

Preface

A Field Programmable Gate Array (FPGA) is a programmable logic device that implements multi-level logic. FPGAs resemble traditional mask-programmed gate arrays by their modular, extensible structure that includes both logic and interconnect, but differ in that their programming is done by end users at their site. No masking steps are required. In this respect, FPGAs resemble PLDs. FPGAs offer low risk, low incremental cost and fast prototyping advantages.

FPGAs are revolutionizing the way systems designers implement logic. By radically reducing the development costs and the turnaround time for implementing thousands of gates of logic, FPGAs provide a new capability that affects the semiconductor industry and the CAE industry. They may also change the way digital systems will be designed in the future.

The Scope of the Book

The field of FPGAs is varied and dynamic. Many different kinds of FPGAs exist, with different programming technologies, different architectures and different software. This book describes the major FPGA architectures available today, covering the three programming technologies that are in use and the major architectures built on those programming technologies. The goal is to introduce the reader to concepts relevant to the entire field of FPGAs using popular devices as examples, without trying to enumerate every commercially-available product.

This book includes discussions of FPGA integrated circuit manufacturing, circuit design and logic design. It describes the way logic and interconnect are implemented in various kinds of FPGAs. It covers particular problems with design for FPGAs and future possibilities for new architectures and software. This book compares CAD for FPGAs with CAD for traditional gate arrays. It describes algorithms for placement, routing and optimization of FPGAs.

The FPGA device descriptions in this book include specifications of capacity and speed. These numbers are continually being debated by manufacturers. This book does not attempt to enter the debate; there was no attempt to reconcile device specifications from different vendors.

FPGA devices and technology are improving rapidly, so specific numbers for gate counts and device speeds may already be obsolete. However, the general concepts, such as programming methods, architectural constraints due to programming technologies, device scaling and preferred design methods will remain relevant long after the specific devices in this book have only historical interest.

Intended Audience

This book is intended to describe all aspects of FPGA design and development. For this reason, it covers a significant amount of material. An extremely detailed discussion of all these areas would make this book prohibitively long. Our intent is to make each section clear to readers with general technical expertise in digital design and design tools. Readers with significant experience in one of these areas may find the discussions superficial in that area, but useful in others.

This book assumes the reader has an understanding of the fundamentals of digital electronics design. Experience designing or using ASIC gate arrays and software will make much of this book much easier to read, since many of the comparisons are with respect to gate arrays.

Potential developers of FPGAs will benefit primarily from the FPGA architecture and software discussion. Electronics systems designers and ASIC users will find this book gives them a background on different types of FPGAs and shows applications of their use, which are useful for deciding when an FPGA is appropriate for an application.

This book may be useful in a university setting where it can be used in support of a comparative FPGA architectures course, as background reading for a digital design course with FPGAs as the target implementation, or as supplemental reading for a Computer-Aided Design course for tools targeted to FPGA design automation.

This book is not intended as a product specification for any integrated circuit or software product.

Organization of the Book

Chapter 1 introduces the FPGA in comparison with other logic implementation techniques. It defines the term FPGA in a form that is both general enough to include all types of devices currently being offered, and specific enough to be a guide for evaluating other devices that may appear. The bulk of chapter 1 is a comparison of FPGAs with mask programmed gate arrays, showing the FPGA advantages and disadvantages.

The following three chapters describe three very different FPGA architectures and software. Chapter 2 describes Xilinx SRAM-based FPGAs, chapter 3 describes Actel antifuse-based FPGAs and chapter 4 describes Altera EEPROM-based FPGAs. The three architectures were chosen because they were the most common FPGAs currently in use and because they are very different in their approaches to field programmable logic. They have different programming technologies, different methods of implementing logic and different interconnection strategies. Each of these chapters includes a discussion of an FPGA family, its architecture, software, and applications.

Each chapter was written by an expert in that particular type of architecture. Each author expresses the concepts in the terminology familiar to developers and users of that architecture. To facilitate comparison of the different FPGAs, the chapters follow a common outline, described in chapter 1. In addition, the reader may use the index as a glossary, as terms with similar meanings are correlated there.

Acknowledgments

I would like to thank all those who contributed directly and indirectly to the success of this book, especially the good folks at Xilinx who allowed me the time for this project, especially Bernie Vonderschmitt, Wes Patterson, Gary Leive and Bill Carter. I wish also to thank the authors of the architecture sections, whose effort and endurance were vital to the completion of this project.

Stephen Trimberger
San Jose, CA

Trademarks

FIELD-PROGRAMMABLE GATE ARRAY TECHNOLOGY

Chapter 1
Introduction

1.1. Logic Implementation Options

An electronic system designer has several options for implementing digital logic. These options include discrete logic devices, often called Small-Scale Integrated circuits, or SSI; programmable devices such as Programmable Array Logic (PALs or PLDs); masked-programmed Gate Arrays or Cell-Based ASICs; and Field Programmable Gate Arrays (FPGAs).

Small amounts of logic can be implemented easily with discrete devices. Each SSI chip contains a few identical gates of a specific type. Designers choose the logic they want from the selection of available chip types. SSI logic is often referred-to as "7400-series," in reference to the widely-used Texas Instruments logic family.

A simple Programmable Logic Device (PLD) is a general-purpose device capable of implementing the logic of tens or hundreds of SSI packages.Pioneered by MMI, a PLD implements logic as wide fan-in two-level sum-of-products of its inputs. It may have optional flip-flops or other logic on the outputs of the sum-of-products array. The best-known PLD is the "22V10", with 22 inputs and 10 outputs, developed by AMD and copied by numerous others. A PLD is programmed by users at their site using inexpensive programming hardware. Power consumption and delay limit the size of the simple sum of products structure to dozens of product terms. Large designs require a multi-level logic implementation.

To implement designs with thousands or tens of thousands of gates, designers can use a Mask Programmed Gate Array (MPGA), commonly called a *gate array*. An MPGA can implement tens of thousands or even hundreds of thousands of gates of logic on a single IC in multi-level logic with wiring between logic stages. An MPGA consists of a base of pre-designed transistors with customized wiring for each design. The wiring is built during the manufacturing process, so each design requires custom masks for the wiring. The mask-making charges make low-volume MPGAs expensive. Typical turnaround times for MPGAs are four to six weeks.

Field Programmable Gate Arrays offer the benefits of both programmable logic arrays and gate arrays. Like MPGAs, FPGAs implement thousands of gates of logic in a single integrated circuit. Like PLDs, FPGAs are programmable by designers at their site,

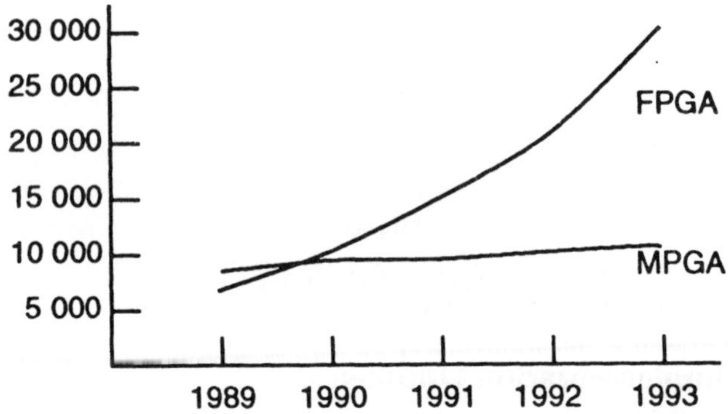

Figure 1.1. Number of Designs Implemented as MPGA Versus FPGA (source: Dataquest, 1991 and Xilinx, 1992)

eliminating the long delays and tooling costs. These advantages have made FPGAs very popular (figure 1.1).

1.2. What is an FPGA?

An FPGA is a general-purpose, multi-level programmable logic device that is customized in the package by the end users. FPGAs are composed of blocks of logic connected with programmable interconnect. The programmable interconnect between blocks allows users to implement multi-level logic, removing many of the size limita-

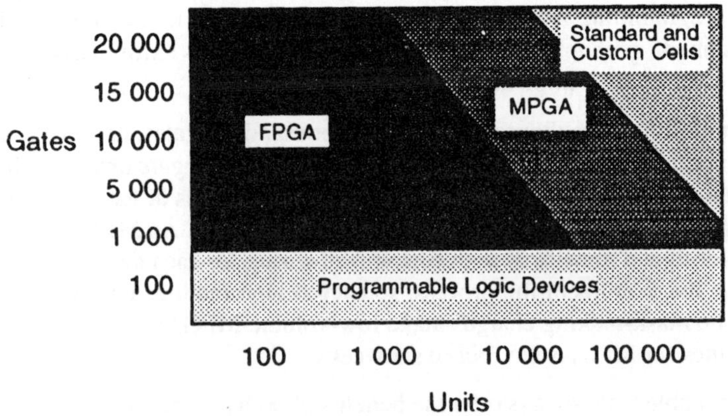

Figure 1.2. Preferred Implementation Options in the Design Space (Source: Xilinx).

tions of the PLD-derived two-level logic structure. This extensible architecture can currently support thousands of gates of logic at system speeds in the tens of megahertz.

The size, structure and number of blocks; and the amount and connectivity of the interconnect vary considerably among FPGA architectures. This difference in architectures is driven by different programming technologies and different target applications of the parts. An architectural organization that works well with a particular programming technology typically does not work with another. The segmentation by programming style and hence architecture is the basis of the taxonomy in figure 1.3. FPGAs fall into four groups: island-style and cellular SRAM-programmed devices; channeled, antifuse-programmed devices; and array-style EPROM or EEPROM-programmed devices.

SRAM-programmed island-style FPGAs include all three Xilinx LCA families, the AT&T Orca and Altera Flex, as well as UTFPGA1 [Chow 1991]. Cellular-style FPGAs include Toshiba, Plessey's ERA, Atmel's (formerly Concurrent Logic) CLi family, the Algotronix CAL, as well as Triptych [Ebeling 1991]. Antifuse-based channelled gate arrays include Actel's ACT-1 and ACT-2, Quicklogic's pASIC and Crosspoint's CP20K Series FPGA. EPROM-programmed array-like devices resemble a collection of PALs with a central interconnection mechanism. Devices of this type are Altera's MAX 5000 and MAX 7000, AMD's Mach and Xilinx's EPLD, among others.

This definition of FPGA is similar to the Complex PLD (CPLD) definition used by the Dataquest market research company. They divide CPLDs into "Programmable Multi-level Devices" (PMDs), which are simple PLD arrays with a programmable

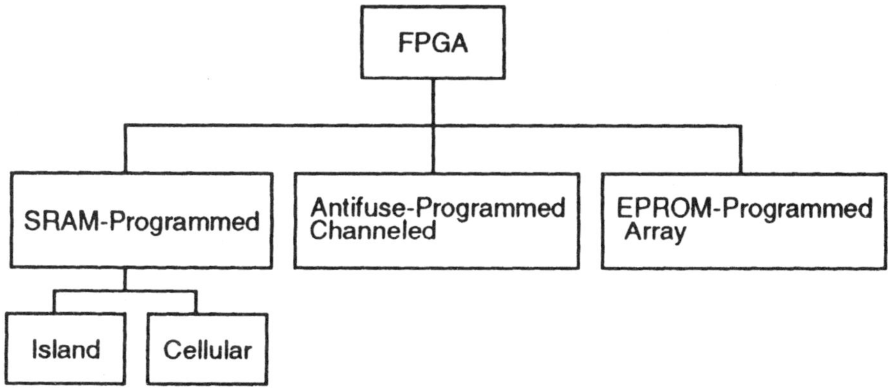

Figure 1.3. Taxonomy of FPGAs.

interconnection structure; and "FPGAs", which consists of all other multi-level field-programmable devices. The Dataquest "FPGA" classification covers those devices in the SRAM-Programmed and Antifuse-Programmed boxes in figure 1.3. In this book, we use the broader definition of FPGAs, which is equivalent to the Dataquest "CPLD" classification.

Currently-available FPGAs implement digital logic, but this is not a fundamental limitation. FPGAs composed of analog blocks with programmable interconnect have been proposed and built [Lee 1991], but they are not commercially available. This book does not describe analog or hybrid FPGAs. The reader is directed to the references for further information on analog FPGAs, also called Field-Programmable Analog Arrays (FPAAs).

1.3. Advantages of FPGAs

Figure 1.4 compares MPGA and FPGA design and manufacturing steps. Design entry and verification are similar for both technologies, but there are significant differences late in the design cycle. Instead of customizing the part by custom manufacturing steps, FPGAs are customized by electrical modification of a packaged part. By eliminating the customization during manufacturing, FPGAs eliminate each design's custom mask-making, test pattern generation, wafer fabrication, packaging and testing. The electrical modification takes milliseconds or minutes, depending on the programming technology and size of the part, compared to weeks for the MPGA steps. FPGA programming is done by simple, inexpensive programming devices.

MPGA	FPGA
System Design	System Design
Logic Design	Logic Design
Place and Route	Place and Route
Timing Simulation	Timing Simulation
Test Pattern Generation	
Mask Making	
Wafer Fabrication	Download / programming
Packaging	
Testing	
System Integration	System Integration

Figure 1.4. Design Steps For MPGA versus FPGA

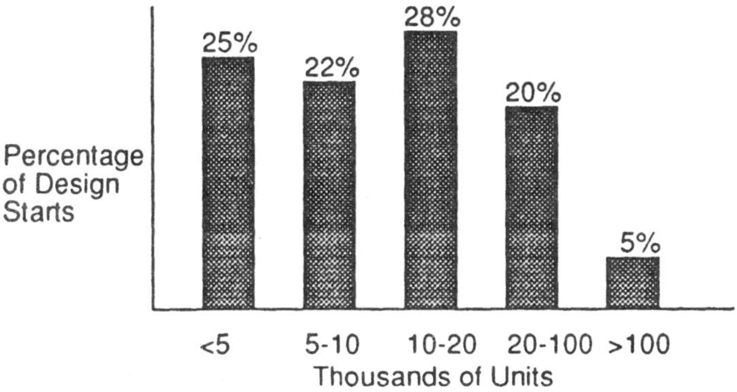

Figure 1.5. MOS Gate Array Design Starts by Unit Volume (source: Dataquest 1991).

Low Tooling Costs

Every design to be implemented in an MPGA requires custom masks to build the custom wiring patterns. Each mask costs several thousand dollars and the cost is amortized over the total number of units manufactured. The more units built, the lower the impact of the masking charges. However, comparatively few designs require more than tens of thousands of units (figure 1.5), so, for most MPGA designs, the masking charges are significant. There is no custom tooling required for an FPGA, so there are no associated tooling costs, making FPGAs cost effective for most logic designs.

Rapid Turnaround

The MPGA manufacturing process takes several weeks from the completion of the design to the delivery of the finished parts. An FPGA can be programmed in minutes by the user of the part. On an FPGA, a modification to correct a design flaw or to address a late specification change can be made quickly and cheaply. Faster design turnaround leads to faster product development and shorter time-to-market for new FPGA products. Reinertsen [1983] determined that in a high-technology environment, a six-month delay in product delivery cut the lifetime profits of a product by thirty-three percent.

Low Risk

The benefits of low initial Non-Recurring Engineering (NRE) charges and rapid turnaround means that a design iteration due to an error incurs neither a large expense nor a long delay. Low costs encourage early system integration and prototyping. The low cost of error also encourages more aggressive logic design, which may yield better performance and more cost effective designs.

Effective Design Verification

Because of substantial NRE costs and manufacturing delays, MPGA users verify their designs by extensive simulation before manufacture. Simulation has an inherent speed/accuracy trade-off: highly accurate simulators are slow, fast simulators are inaccurate. To verify the functionality of the design in a system, large amounts of time must be simulated. Proper verification requires that the environment of the design be simulated as well. Week-long simulation runs are not uncommon. An MPGA design, verified by simulation, may include errors due to inaccuracies or over-simplifications in the simulation model.

FPGAs avoid these problems. Instead of simulating large amounts of time, FPGA users may choose to use in-circuit verification. Designers can implement the design and use a functioning part as a prototype. The prototype operates at full speed and with excellent timing accuracy. A prototype can be inserted into the system to verify functionality of the system as a whole, eliminating a class of system errors early.

Low Testing Costs

All ICs must be tested to verify proper manufacturing and packaging. This test is different for each design. Designs implemented in an MPGA incur three costs associated with testing: on-chip logic to facilitate testing, generation of the test program and testing the parts when manufacturing is complete. FPGAs address all these costs.

Good test programs are hard to write, and schedule pressures tend to abbreviate test program generation for MPGAs. The poor coverage of MPGA test programs allows some bad chips to pass the testing. These defective parts may not be discovered until they fail in a system where the cost of repair is high.

In contrast, the test program for FPGAs is the same for all designs and tests the FPGA for all users of the part. Because there is only one test program, it is reasonable to invest a considerable amount of effort in it, and it can be continually improved over the lifetime of the FPGA. The resulting test program achieves excellent test coverage, leading to high-quality ICs.

In the case of reprogrammable parts, the manufacturer can reprogram all programmable points during testing to verify that the part will work properly after programming. For one-time-programmable parts, the FPGAs generally include test circuitry to catch most failures during manufacture. Manufacturer-supplied hardware and software verify post-programming functionality. Those parts that fail to pass the post-programming test are rejected on the programming device. The percentage of successfully-programmed devices is termed *programming yield.*

The manufacturer's test program verifies that every FPGA will be functional for all possible designs that may be implemented on it. FPGA users are not required to write design-specific tests for their designs. Therefore, designers need not build the testability into the design, eliminating "design for testability" and the design effort and overhead associated with it.

Standard-Product Advantages

New, denser integrated circuit technologies drive microelectronic advances. New processing technologies have finer geometries with smaller transistors and wires. The speed and cost of a chip are related to these dimensions, so a smaller chip is both cheaper and faster.

Moving an MPGA design to a new process incurs additional NRE charges for new masks and test program verification, so it is rarely done. Because FPGAs are standard products, only the FPGA manufacturer incurs the cost of moving the chip to a new process technology. FPGAs on the new process are available to all customers without additional NRE cost. From the point of view of a user, the FPGA manufacturer lowers the price and improves the speed of the parts over time. A user of MPGAs does not get these improvements without paying the additional NRE charges.

Normal variations in the integrated circuit manufacturing process leads to a distribution of performance of integrated circuits. In an MPGA, the customer must design to worst-case process characteristics. The chips that meet "typical" specifications rather than "worst-case" specifications are approximately twenty percent faster. FPGA manufacturers can separate the fast parts from the slow ones in a process called *speed binning*. Slower parts sell for lower price. Faster parts allow designers to design to the high-end of the process variation, giving FPGA users a price/performance trade-off that MPGA designers do not have.

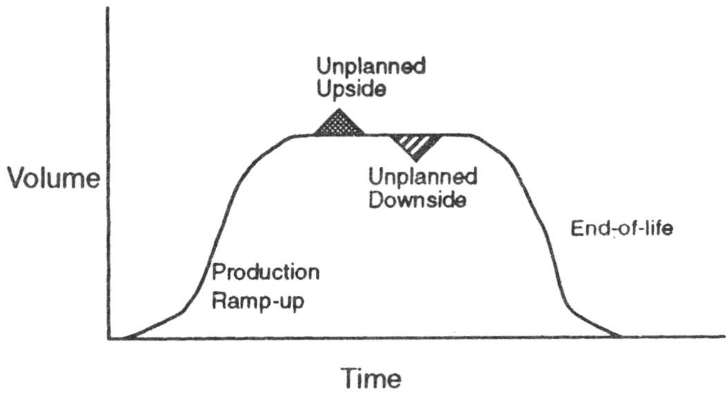

Figure 1.6. Generic Product Life Cycle (source: Xilinx).

Life Cycle Advantages

The life of a product does not end when the design of a chip inside it is finished. Figure 1.6 shows a typical product life cycle plotted as volume versus time. When the design is complete, there is a ramp-up into production. This ramp-up may include a few prototypes or larger pilot manufacturing runs. During production, a design may have periods of increased or decreased sales. At the end of a product lifetime, production tapers off.

MPGA are only cost effective when ordered in volume, and the volume must be decided months in advance of delivery due to long manufacturing cycle times. An MPGA user must maintain a sufficient inventory to handle upturns, and is left with excess inventory should sales fail to meet expectations. At the end of a product lifetime, MPGA users are faced with a last-purchase decision that must be made months in advance of the end-of-product date. If they order too many parts, they are left with unusable parts in inventory, if they order too few, they may not have enough parts to build the last few systems.

The cost-effectiveness of FPGAs in low volume and the flexibility provided by field-programmability provide advantages over all phases of product lifetime. When introducing a product, an FPGA user may order a few parts at a time while testing the design for functionality and the product for market viability. During production, the FPGA user can accommodate rapid changes in sales easily because long lead times are not required. An FPGA user can make product enhancements by shipping an upgraded design on the same FPGA device. This upgrade requires no inventory changes, no new hardware and does not interrupt production.

1.4. Disadvantages of FPGAs

FPGAs have on-chip *programming overhead* circuitry that manages the programming of the part. The area of the programming overhead cannot be used by customers, and lowers the FPGA gate density. The programmable switches and options in an FPGA are larger than the mask programming that can be built in an MPGA. The programmable switches also increase signal delay by adding resistance and capacitance to interconnect paths. As a result, FPGAs are larger and slower than equivalent MPGAs.

Chip Size and Cost

The area penalty for field-programmability is significant. Current FPGAs are about ten times larger for the same gate capacity as the equivalent MPGA, and are correspondingly more expensive on a per-chip basis.

Because of the overhead of field programmability, current-generation FPGAs are limited to tens of thousands of gates of capacity, while the largest MPGAs are hundreds of thousands of gates. For large designs, designers must either split the design into several FPGAs or they must move the design to an MPGA. Multiple-chip partitioning

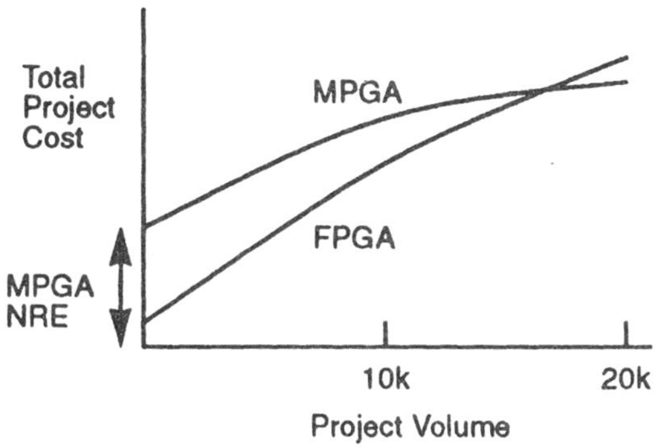

Figure 1.7. Cost Versus Volume for MPGAs vs. FPGAs (source: Xilinx).

for FPGA designs is available, but still relatively immature.

Figure 1.7 shows a cost comparison of MPGAs against FPGAs over a range of volume for the parts. At some point the initial cost savings from the lack of NRE charge with FPGAs is consumed by the increased per-chip costs. That crossover point for a five-thousand-gate part is currently above ten thousand units. Designs with greater volume are more cost effective using an MPGA, despite the greater start-up costs.

Speed of Circuitry

The connection paths in an FPGA are slowed by the programming circuitry. Programmable interconnection points along a wiring path add resistance to the path. All programming points in the interconnect add capacitance to the internal paths. Finally, since more area is required for the same amount of logic, interconnect lines between logic are longer. Longer lines have greater resistance and capacitance, further slowing the resulting circuitry. Current FPGAs are two to three times slower than MPGAs and it is unlikely that they will ever equal MPGA performance on the same manufacturing process.

Design Methodology

FPGAs have been criticized because they are *too easy* to use, thereby encouraging a "try-it-and-see-what-happens" methodology for logic design. If these "sloppy" design practices result in poor-quality designs, the resulting products will be inferior.

1.5. Technology Trends

This section examines the relative performance of FPGAs and MPGAs in the context of advancing process technology.

Density

In ASIC terminology, density is the amount of logic that can fit on a chip. The programming overhead of FPGAs dictates that, for the same amount of logic, FPGAs will always be larger and therefore more expensive than MPGAs. However, many MPGA designs are *pad limited* -- the size of the die is dictated not by the number of gates that can be placed in the area, but by the number of I/O pads that surround those gates. Since I/O pads are placed on the periphery, they scale *linearly* with feature size as a result of improved IC manufacturing processes, while the core of the array scales *quadratically*. Since some fraction of the core of the chip is wasted in a pad-limited MPGA design, the use of that area by FPGAs for field programmability would not increase the area of the resulting part. At some point FPGA and MPGA size for a given number of gates will be dictated by the I/O count, so FPGA and MPGA silicon cost and capacity will be the same.

FPGAs have a fundamental advantage in manufacturing efficiency. Since all parts are identical and all test programs are the same, FPGAs require less equipment and less handling to produce the same number of good parts. This efficiency drives down the cost of FPGAs relative to MPGAs for the same die size.

Speed

Despite being smaller and slower than MPGAs, FPGA size and speed are adequate for most applications. Because field programmability slows down the parts, FPGAs will always be slower than the equivalent MPGA for the same application, but there is reason to believe that FPGAs will reduce the gap in performance.

An application typically has a small amount of logic that dictates its overall speed. Dedicated architectural features of FPGAs can eliminate unneeded programmability in speed-critical paths. These features include high-fan-in decoding logic and special-purpose arithmetic logic. Since there is no extra programming circuitry in the path to slow it down, it runs as fast as custom logic. Since it was designed with the intent to produce a high-speed path, it may even run faster than the mask-programmed general-purpose circuitry in an MPGA.

FPGA manufacturers move their FPGAs to faster processes as they become available, allowing a user to speed up the application simply by buying the faster device. In contrast, an MPGA manufacturer will not automatically upgrade an existing MPGA design to a new process, absorbing the expense of the new mask set. The result is that although the MPGA manufacturer has an improved process, the MPGA user cannot access to it with the old design.

Architecture

Most importantly, though, is the relative immaturity of FPGA architectures. MPGAs trace their lineage to IBM's Master Slice of the late 1960s. They have been refined over the years to an efficient structure that merges logic and interconnect, and serves as an easy target for design automation software. In contrast, the first FPGAs were introduced in 1984, and FPGA architectures are still undergoing significant change. More than any other change, these architectural innovations should serve to close the gap with MPGAs as FPGAs mature.

1.6. Designing for FPGAs

The design flow for FPGAs varies among the different types of FPGAs. This section gives an overview of the problems associated with designing with FPGAs and discusses some common issues. Design tools and methodology will be covered in more detail in the following sections, with emphasis on the design flow for the particular FPGA.

FPGA designs can be made with the same tools and techniques used to design MPGAs (figure 1.8). Schematic entry, logic synthesis and equation-based entry are all available from FPGA manufacturers and CAE vendors. Some FPGAs can be programmed with a very low-level device-dependent design editor, for maximum utilization of the device.

FPGA manufacturers and CAE software vendors supply software for optimizing the logic for an FPGA architecture, for mapping the logic into the FPGA efficiently, and for routing the connections through the configurable interconnect. The result of the implementation step is a programming file that can be loaded into the FPGA or used to control the FPGA programmer to customize the part.

The current generation of ASIC design tools was developed to support MPGAs and standard-cell chips. Since MPGAs and standard cell chips are typically made by interconnecting standard gates in custom patterns, the tools retain this bias. This bias is particularly evident in logic optimization and technology mapping, where the software attempts to optimize the design to a gate-like architecture. This optimization may not produce a good result on an FPGA that does not follow the MPGA model. Different FPGA vendors address this problem differently. Some have chosen to address software compatibility as a fundamental issue, others have chosen to adopt a more MPGA-like "synthesis friendly" FPGA architecture. These software and architecture issues will be addressed independently for each architecture.

Design Migration

FPGAs provide cost-effective design and implementation, while MPGAs provide low-cost volume production. A designer can have the best of both worlds by prototyping a design on an FPGA, then switching to an MPGA for production. An FPGA

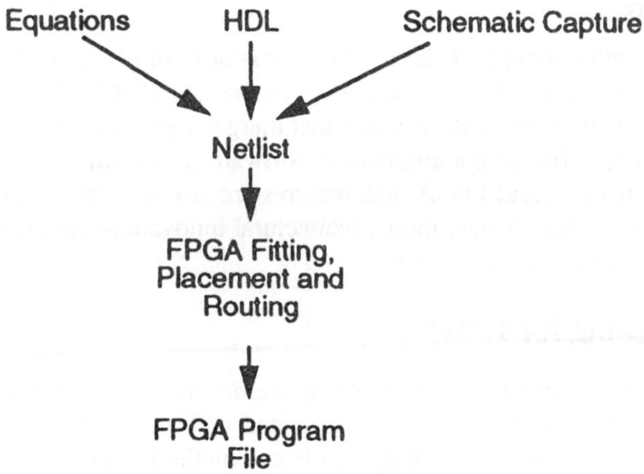

Figure 1.8. FPGA Design System Overview

netlist can be re-targeted to an MPGA by library translation followed by re-verification of all timing paths. The re-verification is difficult if the target MPGA logic does not have the same relative performance as the FPGA logic. Someday, re-targeting software in logic synthesis systems may be able to correct the timing differences, but at present this problem requires manual intervention.

To address this design migration problem, some FPGA vendors offer mask-programmed versions of their field programmable parts with compatible logic structures and delays. The coupling of FPGAs and their mask-programmed equivalent parts give FPGA users the advantages of both implementation methods.

1.7. Outline of Subsequent Chapters

This remainder of this book contains three chapters, each of which addresses a single commercial FPGA architecture and software. Chapter 2 describes the Xilinx SRAM-based architectures, chapter 3 describes the Actel antifuse-based architectures, and chapter 4 describes the Altera EPROM-based architectures.

Each chapter stands as a complete presentation, but the three chapters share a common outline, so readers may make comparisons between the different FPGA architectures and software. The common outline is this:

1. Introduction

This section contains background information on the architecture.

2. Programming Technology

Each of the FPGAs described in this book has a different method of programming. This section describes the manufacturing process and circuit design that provide field programmability.

3. Device Architecture

This section describes the organization of an FPGA, the way it implements logic and the way the logic blocks are connected. This section also addresses density and speed limits of the parts and architectural design trade-offs.

4. Software

Design methodology and flows differ among different FPGA architectures. Architectural decisions may alleviate or emphasize different aspects of the software. This section describes not only the tools needed, but the algorithms those tools use.

5. The Future

Technology trends indicate how the capabilities of these FPGAs will change relative to one another in the future. In addition, this section identifies open areas of research.

6. Design Applications

The advantages and disadvantages of each architecture can be made most apparent with design examples. The example designs in this section show the interaction between the FPGA architecture and the design software.

7. Acknowledgments

8. References

References for each type of FPGA are listed separately for each chapter. A single reference may appear in more than one chapter.

1.8. References

D.G. Reinertsen, "Whodunit? The search for the new-product killers," *Electronic Business*, July 1983.

P.G. Smith, D.G. Reinertsen, *Developing Products in Half the Time*, Van Nostrand Reinhold, New York, 1991.

C. Ebeling, G. Borriello, S.A. Hauck, D. Song, E. Walkup, "TRIPTYCH: A New FPGA Architecture", in *FPGAs*, W. Moore and W. Luk, ed., Abingdon Press, Oxford, UK, 1991.

E.K.F. Lee, P.G. Gulak, "A CMOS Field-Programmable Analog Array", *1991 IEEE International Solid-State Circuits Conference Digest of Technical Papers*, IEEE, 1991.

P. Chow, S.O. Seo, D. Au, T. Choy, B. Fallah, D. Lewis, C. Li, J. Rose, "A 1.2um CMOS FPGA using Cascaded Logic Blocks and Segmented Routing", in *FPGAs*, W. Moore and W. Luk, ed., Abingdon Press, Oxford, UK, 1991.

Chapter 2
SRAM Programmable FPGAs

Steve Trimberger, Xilinx, Inc.

2.1. Introduction

Since their introduction, SRAM-programmable FPGAs have become very popular. Carter [1986], Hsieh [1987, 1990], Kean [1989], Furtek [1990], Hastie [1990], Kawana [1990], Muroga [1991], Ebeling [1991], Chow [1991], Hauck [1992], Hill and Britton [Hill 1992][Britton 1993] and Cliff [1993] have all proposed SRAM-programmable FPGAs.

This chapter focuses on the three Xilinx families of FPGAs [Carter 1986] [Hsieh 1987] [Hsieh 1990] as representatives of the class of SRAM-programmable FPGAs with mature software. These three FPGA families share a common structure: an array of configurable logic blocks surrounded by configurable interconnect. The three families differ in the details of the logic and interconnect structures. Members of a family have identical block and wiring structure, but differ in the size of the array. This chapter begins with an overview of the programming technology, it covers device architectures for the three devices, software and design applications.

2.2. Programming Technology

SRAM Programming

An SRAM-programmable FPGA is programmed by loading *configuration memory cells* from an external source. The configuration memory cells control the logic and interconnect that perform the application function of the FPGA. There is no separate RAM area on the chip, the memory cells are distributed among the logic they control.

The configuration memory is written only once for each application so, unlike commercial static RAM memory chips, high-speed read and write is not important. Stability and density are primary concerns. Figure 2.2.1a shows the CMOS five-transistor memory cell used in Xilinx FPGAs. The Read/Write pass transistor (R/W), is used both to load the cell and to read back the programming. During normal operation it is off, and the cell holds its programming.

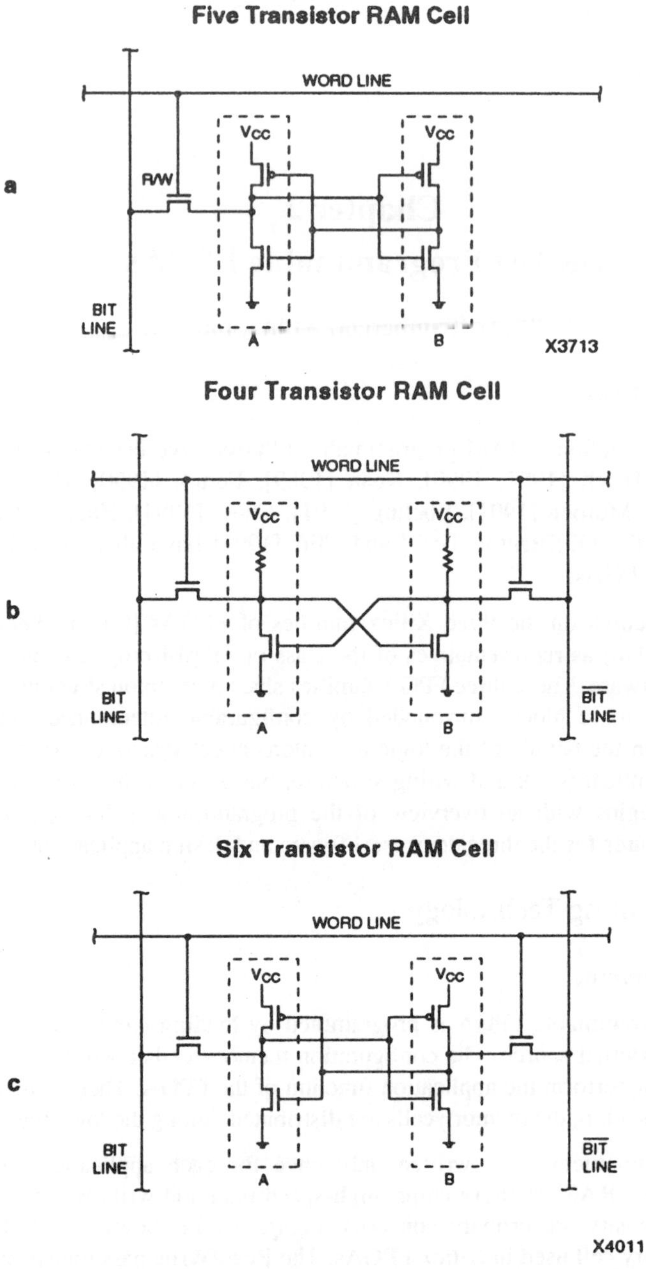

Figure 2.2.1. a) Xilinx Five-Transistor Configuration Memory Cell. b) Four-Transistor Memory Cell. c) Six-Transistor Memory Cell.

The six-transistor memory cell used in CMOS memories uses the data bit in both the true and complement form (figure 2.2.1c), providing fast read and write times at the cost of another transistor. The four-transistor memory cell often used in high-density SRAMs (figure 2.2.1b) has polysilicon resistors instead of the P-channel pullup transistors. These giga-ohm resistive paths increase sensitivity to soft errors such as alpha-particle upsets. The internal signals in the memory cell in figure 2.2.1a are always connected to one of the two power supplies with a low-resistance path, so the cell is very stable. Alpha particle sensitivity tests of the five-transistor memory cell show the expected mean time between failure due to alpha particle upsets to be nearly one million years [Lyons 1985] [Xilinx SMTO 1991].

Advantages and Disadvantages of SRAM Programming

Volatility

SRAM programming has an obvious drawback -- volatility. When the power is turned off, the IC loses its programming, so an SRAM FPGA must be reprogrammed each time power is applied. SRAM FPGAs include logic to sense power-on and to automatically initialize themselves, providing "virtual non-volatility," provided the application can wait the 2ms-30ms required to program the FPGA. This is usually not a problem because system start-up times are usually longer than this. For systems that need active logic during power-up, the initialization time for an SRAM-programmed part must be considered in the power-up sequence.

External Memory

A related disadvantage of SRAM programming is that it requires an external memory for permanent storage of the program. Although multiple FPGAs can share a single external memory, this multiple-chip solution may be inappropriate where board space is crucial. However, many systems already have significant amounts of storage in initialization microcode, bootstrap PROMs or disks. Many users of reprogrammable FPGAs are able to share the permanent memory used to initialize other components of the system.

Reprogrammability

The disadvantage of volatility provides the advantage of reprogrammability. Reprogrammability makes SRAM-programmable FPGAs ideal for prototype development. Since the FPGA can be reprogrammed without cost, a designer can load a design into the part, try it at-speed in the system and debug the design. If necessary, a modified version of the design can be loaded into the FPGA and tried in the system without removing the chip. One-time-programmable FPGAs require that the defective prototype part be removed and a new part programmed and inserted into the system to make the change.

A reprogrammable FPGA can be time-shared to replace logic amounting to many times its maximum capacity. If a system can be divided into pieces that are not required simultaneously, the pieces can be designed into separate configurations of the FPGA, and the FPGA time-shared between them by repeated reprogramming.

A common use of reprogrammability is for board-level test. As part of the test sequence, the FPGA is configured as a test generator/checker to verify the board or components on the board. After board-level test is finished, the FPGA is re-configured with the application logic. This design saves the additional hardware that would have been needed for system-level test. Other uses allow a system to interface to multiple different external devices, selectable by the programming of the FPGA.

A system built with reprogrammable logic can be updated after delivery to a customer by modifying the programming of the FPGA. The new programming bitstream can come from a new PROM or from memory in some other part of the system. This feature has been used for soft-hardware field upgrades, replacing the FPGA programming from a floppy disk or even from a remote computer over a modem. Applications of reprogramming are discussed in more detail in the Applications section of this chapter.

Quality

Indirectly, reprogrammability leads to very high quality parts because each part can be fully tested at the factory without destroying it. Every programming point and every path is tested. SRAM FPGA tests cover all the typical stuck-at faults as well as many other pattern faults that ASIC test generators might miss. These faults include stuck-open faults and bridging faults that result from layout proximity. SRAM FPGAs can also be tested for speed and binned accordingly, so users can choose only the fastest parts for performance-critical applications.

Programming yield for SRAM FPGAs is always 100%. There is no separate programming step, and no removal and re-insertion for programming. The handling associated with programming antifuse and EPROM-programmed devices damages the package pins, causing problems for board assembly.

Process Leadership

The SRAM process used to build FPGAs is the same CMOS process used to make ASICs and is very similar to the process used for CMOS memories, so SRAM-programmable FPGAs are among the first logic products to take advantage of process improvements driven by semiconductor memories. Since improved processes are both denser and faster than older ones, those advantages apply to SRAM-programmed FPGAs as well.

Low Power

SRAM-based FPGAs implement logic in static gates, so SRAM programmable FPGAs have low power consumption even for very large amounts of logic and have zero standby current. In contrast, EPLD-style FPGAs have passive pullup and sense-amplifier circuitry that leads to prohibitively-high static power dissipation for high-capacity or high-speed parts.

2.3. Device Architecture

All three Xilinx FPGA families consist of an array of Configurable Logic Blocks (CLBs) embedded in a configurable interconnect structure and surrounded by configurable I/O blocks (figure 2.3.1). Each family architecture was driven by a different set of assumptions and a different model of use. These different design pressures led to different block, I/O and wiring architectures, as well as other unique features of each FPGA. Family members differ in their number of blocks and I/Os, with CLB array sizes ranging from 8x8 to 24x24 blocks. These chips support designs in excess of ten thousand gates, with system clock speeds in the tens of megahertz.

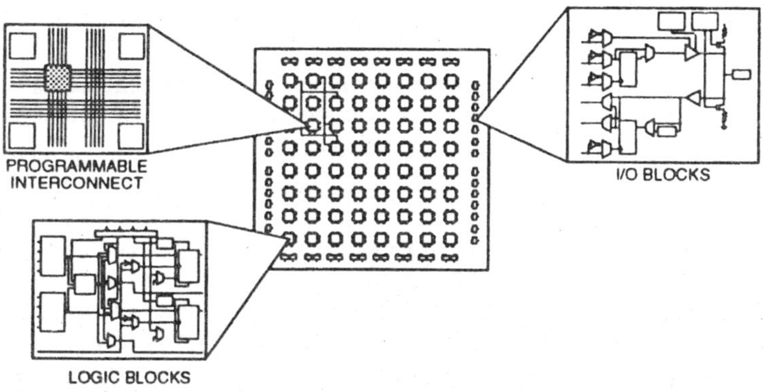

Figure 2.3.1. Island-Style SRAM-Programmable FPGA Architecture.

Simple SRAM-Programmable FPGA Architecture

This section contains a bottom-up description of a simple SRAM-programmable FPGA followed by a discussion of architectural trade-offs. Rather than try to describe a commercial FPGA architecture, the intent is to give the reader an understanding of the underlying programming technology and design methodology without the detail of a particular implementation. Following sections describe commercial architectures and how they address the design trade-offs.

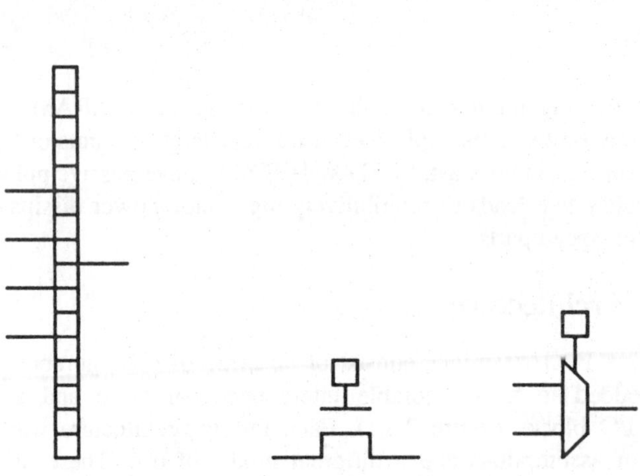

Figure 2.3.2. Three Important Pieces. a) Lookup Table. b) Pip. c) Multiplexer Controlled by a Configuration Memory Cell.

Building Blocks

Lookup Table. Figure 2.3.2a shows a four-input *lookup table* (abbreviated *LUT*) or *function generator*, a basic unit of configurable logic. A lookup table implements combinational logic as a 2^nx 1 memory composed of configuration memory cells. The memory is used as a lookup table, addressed by the *n* inputs. A lookup table can implement any of the 2^{2^n} functions of its inputs. When the FPGA is programmed, the memory is loaded with the bit pattern corresponding to the truth table of the function [Horowitz and Hill 1989]. For example, if all configuration bits in the 16-bit lookup table in figure 2.3.2a are 0 except the high-order bit (address 15), the output of the lookup table will be zero unless all inputs are high (binary address 1111): a 4-input AND. To implement functions of fewer inputs, the unused inputs are held low or the subset of the truth table entries is duplicated to make the output the same regardless of the value of the unused input.

The inputs to the lookup table are logically equivalent -- any signal can be connected to any input pin of the lookup table. Changing the pin to which a signal is connected requires a straightforward rearrangement of the bits in the lookup table. All functions of a lookup table have the same timing: the access time of the memory. Placement and routing software can take advantage of the logically-equivalent pins, choosing which input connects to which pin to optimize routing.

Programmable Interconnect Point. The second building block is called a *programmable interconnect point, (pip)*. Some authors use the term "configurable interconnect point" (cip). Pips control the connection of wiring segments in the programmable interconnect. The pip, shown schematically in figure 2.3.2b, is a pass transistor controlled by a configuration memory cell. Wire segments on each side of the transistor are connected or not depending on the value in the memory cell. A pip is the basic unit of configurable wiring.

The pip switches the signal with a pass transistor, not a full CMOS transmission gate. The transmission gate requires three more transistors: one for the transmission gate and two for the inverter that negates the control signal. A full CMOS transmission gate requires two transistors connected to each wiring segment for each pip, adding diffusion capacitance to the segment, slowing down the interconnect. The drawback of the single pass transistor design is *threshold drop*: signals in the interconnect are not pulled all the way up to the high voltage, so all logic inputs must either dissipate power due to lowered high-voltage on the P-channel pullup, drive the P-channel destination with a restored signal, or include a CMOS P-channel pullup with a lower threshold to restore the signal logic.

Multiplexer. The third building block is a multiplexer controlled by a configuration memory cell (figure 2.3.2c). The multiplexer is a special-case, one-directional routing structure. It may be of any width, with more configuration bits for wider multiplexers. Switches built with multiplexers reduce the number of memory cells required for controlling the switch, giving an area savings for large switches.

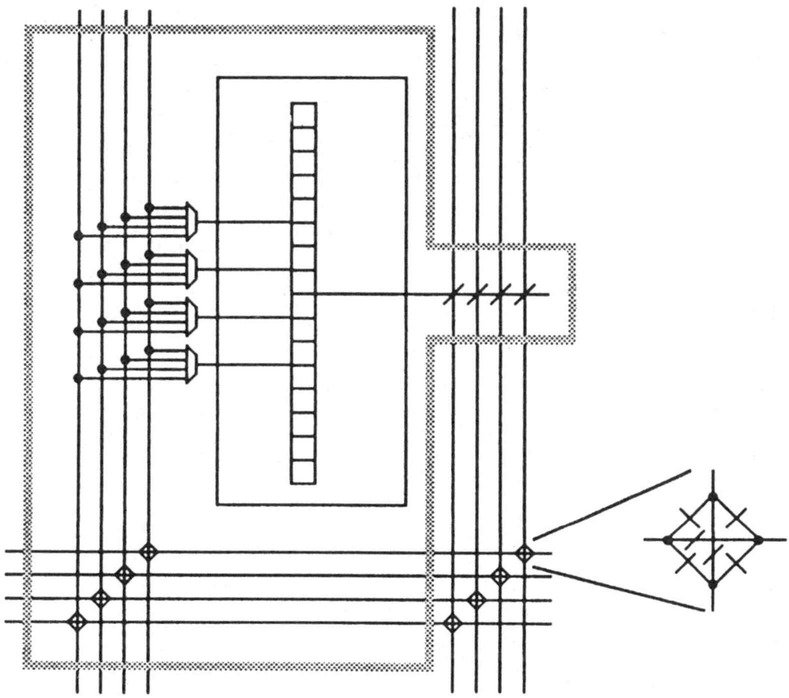

Figure 2.3.3. An FPGA Tile.

The Logic Block

Figure 2.3.3 shows the building blocks from figure 2.3.2 combined into a single *tile*, shown outlined, composed of a *configurable logic block* (*CLB*) and wiring. The CLB in figure 2.3.3 contains a single four-input lookup table. The CLB is surrounded by *wiring channels*. Each wiring channel contains several wiring *segments*. Segments in the left side channel have multiplexers on them to make connections to the CLB. The CLB output is connected to the wiring segments at the right through pips. For simplicity, pips are shown as diagonal lines at the intersections of the segments in the channels and between segments in the switchbox. If the block is unused, the output pips are all turned off, so the block does not drive its signal onto any segment, and the segment may be used to connect other signals on the chip.

Where the wiring channels intersect, their connections are made by a pattern of pips in a *switchbox*. For routability considerations, a crossbar switch in the switchbox is ideal, but a full crossbar switch connecting n wires requires n^2 pips, so it is prohibitively large. In practice, the switchbox contains only a few of the possible connections between segments in intersecting channels. A signal may be routed through pips in switchboxes to other blocks or back to the same block. Each segment in the switchbox in figure 2.3.3 has three pips, one to a segment on each of the other three sides of the switchbox. The number and pattern of connections in the switchbox has a significant effect on the routability and performance of the FPGA, as will be discussed later.

The tile in figure 2.3.3 can implement sequential logic as well as combinational logic. To build a latch, the lookup table is configured as shown in figure 2.3.4, and the output is routed back to the input. The first and second inputs to the lookup table are configured for the reset and set signals. A clocked latch can use the fourth input for the clock. Latches can be combined to implement other kinds of sequential elements, such as D-type or T-type flip flops.

The Chip

Figure 2.3.5 shows a chip composed by building an array of CLB tiles, then surrounding the array with configurable *I/O Blocks* (*IOBs*). An IOB allows signals to be driven off-chip or optionally brought onto the chip onto interconnect segments. The IOB may perform other functions, such as three-state outputs and registering incoming or out-going signals.

Different family members in a common architecture are constructed by assembling different-sized arrays of the tile and surrounding that array with different numbers of IOBs.

The FPGA must include circuitry to load the *configuration bitstream* or *program* into configuration memory where the memory cells produce logic functions in lookup tables and control interconnect in pips and multiplexers. Typically, the bits in the

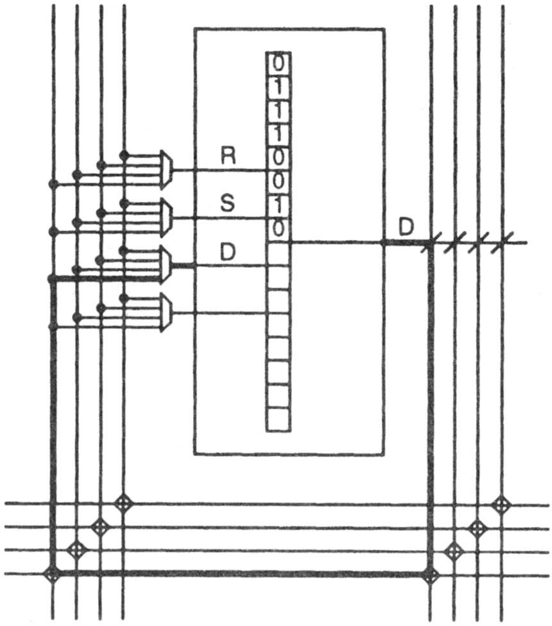

Figure 2.3.4. The Tile Configured as a Latch.

program are loaded serially, and the programming of the FPGA can be considered to be a single, long programming word.

Design Trade-offs

Density and Speed

Density is the amount of usable logic per unit of chip area. Density has two components: the size of the structures that implement logic and the logic capacity of those structures. The size of components in an SRAM FPGA correlates fairly well with the number of memory cells required to build and control them. The capacity of the structures depends on how well software can map logic from typical designs into the structures the FPGA implements. More general structures are preferred, since more logic can be put into them and they require simpler design automation software. However, more general connections usually require more controlling logic, so they take more area.

The area dedicated to interconnect does not contribute to logic capacity, so an FPGA with minimal interconnect may appear to have excellent logic density. However, much of the logic will be inaccessible to designs built on the FPGA because of the

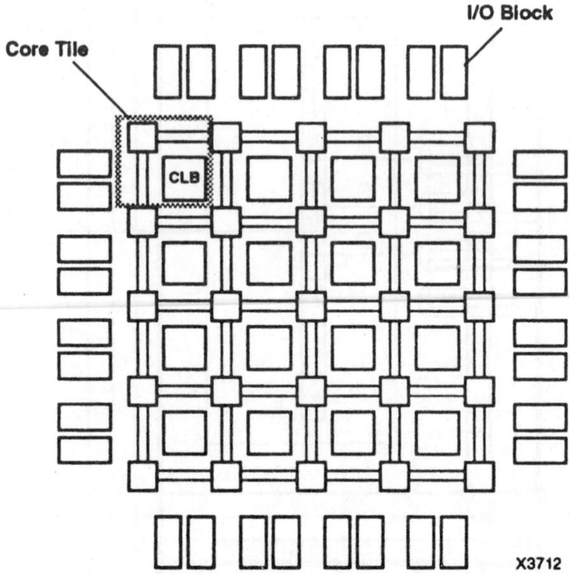

Figure 2.3.5. An FPGA as an Array of Tiles.

inadequate wiring resources. As a result, the wiring-poor FPGA will have comparatively little *usable* logic.

The *speed* of an FPGA is a measure of the delay required to implement a function and to propagate signals to neighboring functions. The speed of the logic part of the block is the sum of the delays from the input selection multiplexers, the lookup table and the output drivers. For complex blocks, the delays of different paths through the block may be different.

The fraction of delay incurred due to interconnect in an FPGA is significantly greater than that in an MPGA because configurable interconnect is inherently much slower than mask-programmed interconnect. In an MPGA, wiring is implemented on metal runs that have low capacitance and low resistance. While an SRAM-programmable FPGA also uses low-resistance metal for interconnect, a signal passes through a transistor at each pip (figure 2.3.6). The channel resistance of the transistor that makes up the pip and the capacitance due the source/drain diffusions of all pips on the segment determine the speed that signals propagate through the interconnect. In figure 2.3.6, a signal from output A to input B sees the resistance of the pip after the output buffer, the loading on segment s4, the resistance of the pip to segment s1, another segment load, and finally the input multiplexer. The equivalent schematic of the path is shown in figure 2.3.6b, where C_s is the segment loading capacitance, equal to the capacitance of all the pips on the segment plus the capacitance of the metal

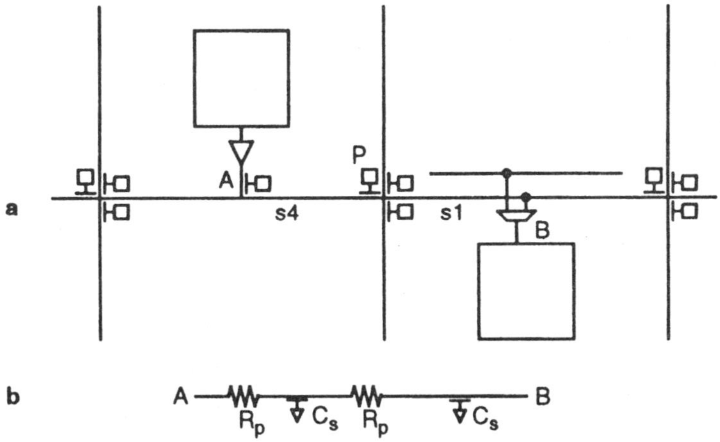

Figure 2.3.6. Interconnect Segment Detail. a) Architecture-Level. b) Electrical Equivalent.

segment, and R_p is the resistance of a pip. Signals that travel longer distances have additional resistance and capacitance stages.

It is the task of both the architecture and the software to limit signal delay and to provide buffering for long wires. The architecture may provide buffering in the interconnect, or the routing software may be required to route signals that need buffering through unused CLBs. Both options have been employed in SRAM FPGAs, and they will be discussed with the individual FPGA architectures.

Speed and density must be traded off against one another. Faster output buffers are larger. CLBs with a large number of inputs can make high-fanin logic faster, but each input multiplexer costs area and hence reduces density.

Size versus Routability

An architecture with more pips provides more options to the router, simplifying routing, but each pip includes a memory cell, adding measurably to the size of the FPGA. Rose and Brown [1991] investigated the routability of lookup-table-based FPGAs under differing numbers of pips in the switchbox and in the routing channel. They placed and routed a set of designs on a simple FPGA architecture, varying the number of tracks in the channels, the number of pips from CLBs to tracks and the number of pips between tracks in switch boxes. They report the number of tracks in excess of the channel density required to route the set of designs over a range of pip distributions. Their results show that, for a high probability of complete routing, the architecture must have good connectivity in the channel, with at least seventy percent of the channel segments connectable to the CLB; but connections in the switchbox may be sparse, with as few as three or four connections from each segment in a switchbox.

Speed versus Routability

Every segment of configurable interconnect is capacitively loaded by the pass transistors of all potential connections that can be made from that segment, as shown in figure 2.3.6. When a signal is routed through a pass transistor to another segment, the net includes the resistance of the transistor. The delay of a net is the time required to charge or discharge the distributed capacitance of all segments through the resistances of all pips and the CLB output driver. The additional delay on a net from adding a routing segment grows quadratically with the number of pips on the net, due to the distributed resistance of the pips and the capacitance of the segments. This is the same delay behavior observed in a lossy transmission line. To improve performance, the resistance of a pip can be reduced by making the pass transistor wider. However, wider transistors load the segment with more capacitance. In practice, interconnect delays are about the same as the block delays, but for pathological cases, they can be much larger.

Because signal delay is more dependent on the number of pips through which a signal is routed than on the distance covered, FPGAs typically have interconnect wiring segments in a variety of lengths. This can be done by replacing some "straight-ahead" pips (such as pip *P* in figure 2.3.6a) with a wired metal connections. This saves the area of the pips, and eliminates the series resistance of the pips along the wiring track. The capacitance of a segment is larger because the one segment still has all the branching pips. By eliminating many switchbox pips, we can make a channel with wiring segments of many different lengths. The result has been called a "segmented routing channel" [Greene 1990].

If a signal travels the full length of a multiple-length segment in a segmented routing channel, it reaches its destination in less time than it would if it were routed on single-length lines. However, if the signal is not required to travel the whole length of the segment, the delay may be greater due to the larger capacitance of the longer segment. Further, if much of the long segmented routes are wasted, the overall routability of the chip is degraded. Therefore segmented channels require more tracks than channels with only single-length lines.

Another architectural option is to include direct connections from the output of a block to the inputs of nearest-neighbor blocks, bypassing both the series resistance of the pips and the capacitive loading of the segments. These connections provide high-speed paths for simple connections. This technique has also been called "cascade." Finally, an FPGA may also have longer-distance connections or special facilities for long-distance or high-fanout connections.

Block Size and Structure

A lookup table is not the only way to implement logic. For example, instead of a lookup table, we can implement logic with a non-programmable four-input NAND gate. The NAND gate requires a fraction of the transistors of the lookup table and runs somewhat faster, although several NAND gates may be required to implement a function that would fit into a single lookup table. For more generality, one could implement a few different functions of the inputs in a logic block and select the one desired function with a programmable multiplexer. This solution still requires some control circuitry: memory cells to control the selection of the function of the block from the set of implemented functions.

Simple blocks, such as NAND gates or two-input lookup tables, are conceptually elegant, and design software can usually use them efficiently. Further, a large block may not always be fully utilized, leading to lower logic capacity.

When considering size and performance trade-offs, one must consider not only the logic element, but also the interconnect segments, pips and multiplexers required to connect the elements. Because the lookup table can implement any function of its inputs, a single lookup table replaces several simple gates as well as the delay and area-intensive interconnect between them. In practice, implementing logic in a lookup table produces a smaller, faster FPGA because the lookup table efficiently combines versatile logic implementation with the programming overhead of the memory cells.

This trade-off can be illustrated with a consideration of lookup table widths. A four-input lookup table has sixteen memory cells, a three-input lookup table has only eight. If the logic being implemented in 4-input lookup tables breaks naturally into three-input gates, half of every 4-input lookup table area will be wasted, leading to an inefficient implementation on the FPGA. On the other hand, if the logic tends to break into four-input pieces and we implement it in 3-input lookup tables, each 4-input function will need three three-input lookup tables plus the associated interconnect. The resulting logic is not only slower, but also larger than the four-input version.

Rose, et. al. [1990] investigated a range of lookup table sizes and their effect on the overall chip area, including both CLB area and routing area. They mapped logic from a number of designs into lookup tables and estimated the size of the chips that would have been required to implement those designs. Their estimate used total memory cell count and wiring cost estimates. They performed their analysis for a range of lookup table sizes and programming cell sizes. Their results show that a three-input or four-input lookup table gives the best density for a wide range of programming cell sizes. Larger lookup table sizes were preferred for a high-speed architecture.

This type of trade-off analysis applies to sequential logic elements as well as combinational logic. Figure 2.3.7 shows a dedicated flip-flop in the logic block with a control multiplexer to select either the lookup table or the flip-flop output. The dedicated flip-flop, when used, saves two lookup tables and their associated wiring.

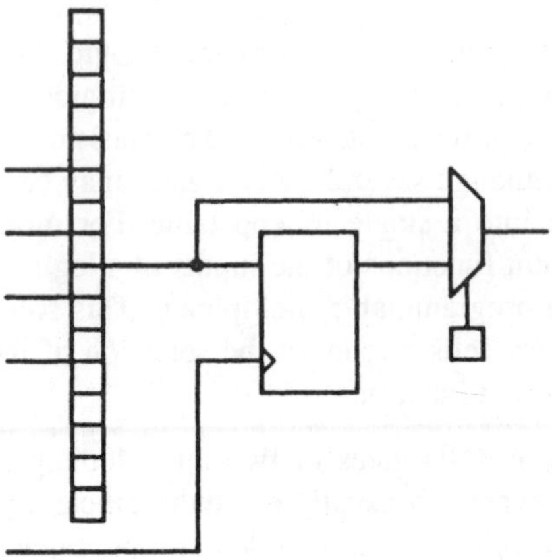

Figure 2.3.7. A CLB with Registered Output.

Although the flip-flop area is wasted when the combinational output is desired, the area savings over all designs for using the flip-flop exceeds the waste when it is unused [Rose 1990]. In addition, the dedicated flip-flop provides predictable setup time, hold time and metastability parameters. If a latch is built from a lookup table and interconnect, as shown in figure 2.3.4, the D connection from the output to its non-feedback destination loads the feedback path, making the timing of the latch dependent on its wiring. Buffering the signal requires an additional lookup table, doubling the size of the latch.

The determination of the desirability of a dedicated function in the FPGA architecture can be generalized into a pair of design rules.

1. A dedicated function improves the density of an FPGA if, over all designs, the area wasted when the function is unused is less than the total area saved when it is used.

2. A dedicated function improves the performance of an FPGA if, over all designs, the delay along the critical path added by the feature is less than the delay reduction along the critical path due to the feature.

The value of a dedicated feature is also dependent on the ability of design software to use it and on the "general applicability" of the feature. A feature that is unused by design software can not improve the quality of implementation. A feature that improves the FPGA for some designs and degrades it for others will be of limited value.

Capacity Estimation

FPGAs have three kinds of resources: logic, I/O and routing. To determine if a design fits into a particular FPGA, the design must fit within all three resource limits. The difficulty of this estimation is a function of the architecture and of the software used for mapping the logic into the FPGA. FPGA logic and interconnect capacity are difficult to estimate. Traditional measures of gate count and product terms are not accurate estimates of lookup-table capacity. Two designs that appear to be of equal size in terms of MPGA gate count or number of PLD product terms may use CLBs with different efficiency, requiring very different numbers of CLBs. Logic optimization algorithms may also significantly change the size and performance of the design.

Complex blocks implement complex functions efficiently, but when the function to be implemented does not fit into the block efficiently, some fraction of the block is unusable, and is wasted. The wasted fractions of blocks cause a gap between the peak capacity of the FPGA and the capacity in a given application. An accurate capacity and performance estimate requires that the design be mapped into the FPGA. Fortunately, fast mapping heuristics can give a good estimate of logic capacity.

Routing requirements are more difficult to estimate. The problem of statistical wirability estimation has been addressed by Heller [1978], Donath [1979] and ElGamal [1981], but the techniques and results are not accurate enough for capacity estimation. MPGA designs address this problem by providing significantly more interconnect than is needed by most designs. This solution in impractical in FPGAs because unused FPGA interconnect degrades performance and density too severely. FPGA design systems include high-speed placement and routing for routability estimation and timing-driven routing to meet delay requirements.

The Xilinx XC2000 Architecture

The Xilinx XC2000 family [Carter 1986] was the first commercially-available FPGA. Introduced in 1985 and still used today, the XC2000 architecture was developed without supporting software to verify logic density or capacity. The block structure was derived from a general understanding of the way logic is decomposed in typical applications and from manual implementations of existing MPGA designs. A crucial concern in the design of the XC2000 family was to build a chip that was small enough to be manufacturable with the IC process available at the time. Therefore, a smaller, slower cell was preferred to a larger, faster cell.

The XC2000 CLB

The XC2000 CLB combinational logic section (figure 2.3.8) consists of two three-input lookup tables producing the F and G signals. The two lookup table outputs can be multiplexed together to produce any function of the four inputs on both outputs. Hill [1991] describes this arrangement as a single four-input lookup table with two

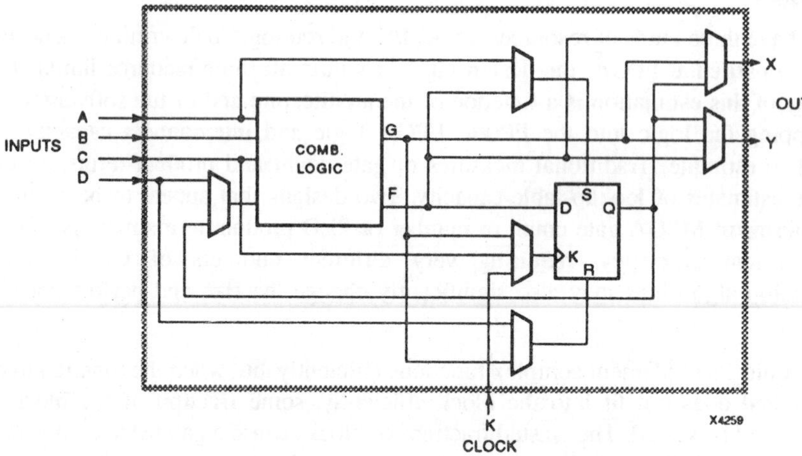

Figure 2.3.8. The XC2000 CLB.

outputs. The CLB includes a single storage element that can be configured as an edge-sensitive D-type flip-flop or as a level sensitive D-type latch. The data input to the storage element comes from the output of the F lookup table. The clock input can come from the G lookup table, the C input to the CLB or from a separate clock input, K.

Either of the CLB outputs can be configured to be the result of the F lookup table, the G lookup table or the sequential result, Q. The output of the flip-flop can be recycled directly to the inputs of the lookup table, providing an efficient method of generating state machines and counters.

The designers of XC2000 architecture determined that four-input lookup tables efficiently implemented most logic and that a dedicated flip-flop was generally applicable. This implied an architecture much like figure 2.3.7. However, they did not want to lose half the lookup table for related functions of three or fewer inputs, such as the sum and carry in a full adder. Therefore the XC2000 CLB has two outputs. For a full adder, it can implement the sum in one lookup table and the carry in the other. The sum can be stored in the CLB flip-flop to build an accumulator efficiently.

The XC2000 IO Block

Figure 2.3.9 shows the XC2000 I/O block structure. All chip outputs can be three-stated and bidirectional. The three-state control can be fixed in the configuration bitstream to make the block input-only or output-only, or it can come from a signal in the FPGA interconnect, so on-chip logic can control the direction of the I/O pads. The input signal can be latched in the IOB, reducing hold times for latched inputs that would otherwise have to be wired to a CLB flip-flop.

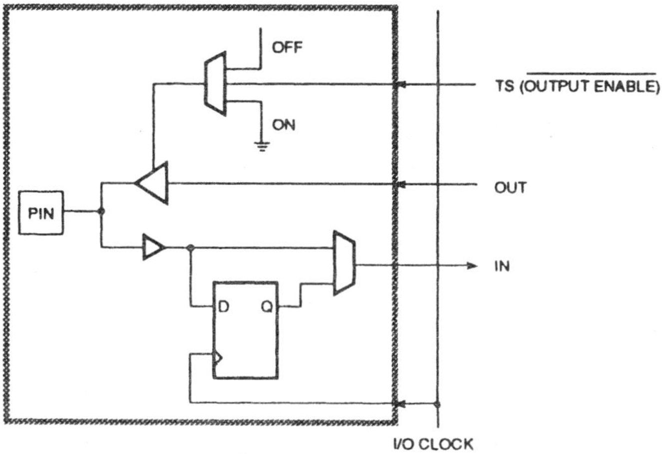

Figure 2.3.9. The XC2000 IOB.

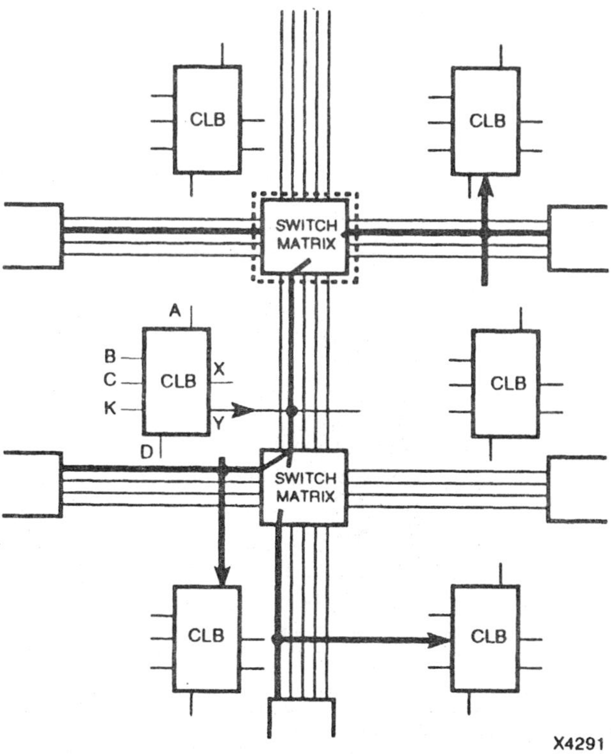

Figure 2.3.10. XC2000 Interconnect Structure.

Wiring Architecture

The XC2000 includes four horizontal and five vertical *general-purpose* interconnect segments between switchboxes in the array (figure 2.3.10). The switchbox pips connect the segments in pairs, with the fifth vertical segment making connections to some of the adjoining segments (figure 2.3.11). The segments can be grouped into channels in which each segment is part of a track.

A net routed on general interconnect shows a distributed RC delay. Sizing all drivers to overcome the worst-case load across the chip is impractical. Instead, the XC2000 provides *repowering buffers* in the interconnect to speed up long-distance connections. The array of tiles is divided into nine sections in a grid arrangement (figure 2.3.12) with general interconnect signals re-powered every time the signal crosses from one section into the next. Local signals within a grid section are not repowered.

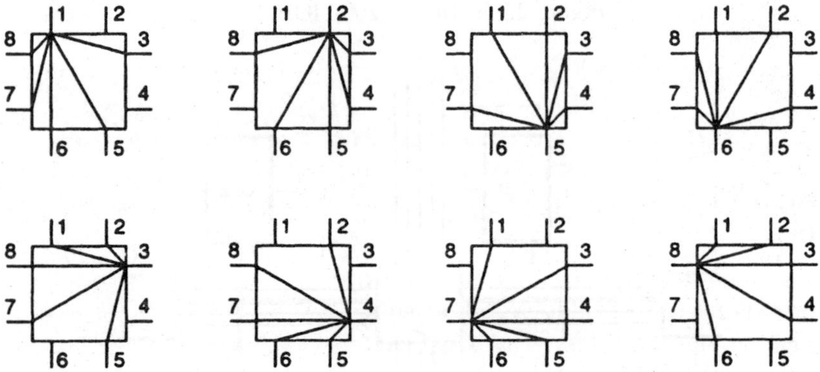

Figure 2.3.11. XC2000 Switchbox Connections.

The XC2000 wiring includes *direct connections* to horizontally and vertically adjacent blocks. These connections are at the ends of the bold lines in figure 2.3.13. A direct connection provides a high-speed dedicated interconnect path to two adjacent CLBs through a single multiplexer input, avoiding the general-purpose interconnect. It is useful for those cases where a signal has a single speed-critical destination and that destination can be placed next to the source. If this ideal placement cannot be made, the connection can still be made by routing through general-purpose interconnect.

The XC2000 wiring includes two vertical *long lines* and one horizontal *long line*. A long line is a single metal segment that spans the entire width or height of the array of tiles, bypassing all switchboxes. Signals are buffered onto the long lines, so that they

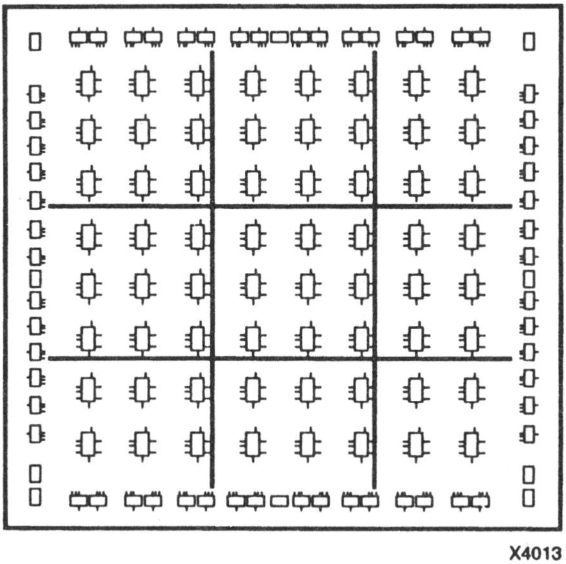

Figure 2.3.12. XC2000 Repowering Buffer Pattern.

can be distributed over a long distance and to a large number of destinations quickly and with low skew.

To further support high-fanout, low-skew signals, the XC2000 FPGA includes two high-drive buffers with dedicated *global* interconnect to all CLBs in the whole chip. These buffers are intended to be used for clocks or other synchronizing signals. This global interconnect reduces the need for a user of the FPGA to do complicated special-purpose clock routing or simulation, as is required in an MPGA to guarantee low-skew clocking paths.

Block-to-Interconnect Connections

The connections to the CLB are distributed around the four sides of the block, with inputs on the bottom, left and top; and outputs on the right (figure 2.3.14). This asymmetry encourages the implementation of a design where signals flow from left to right and top to bottom across the chip. Input multiplexers in the figure, shown as boxes, appear as independent pips to improve the clarity of the drawing. Because equivalent lookup table inputs come from different channels, the router has the flexibility to swap pins.

The outputs of the CLB can connect to only half the segments in the channel. Since CLB inputs can come from any segment, and since signals can switch tracks in the switchboxes, the limited number of output connections is not a serious routability

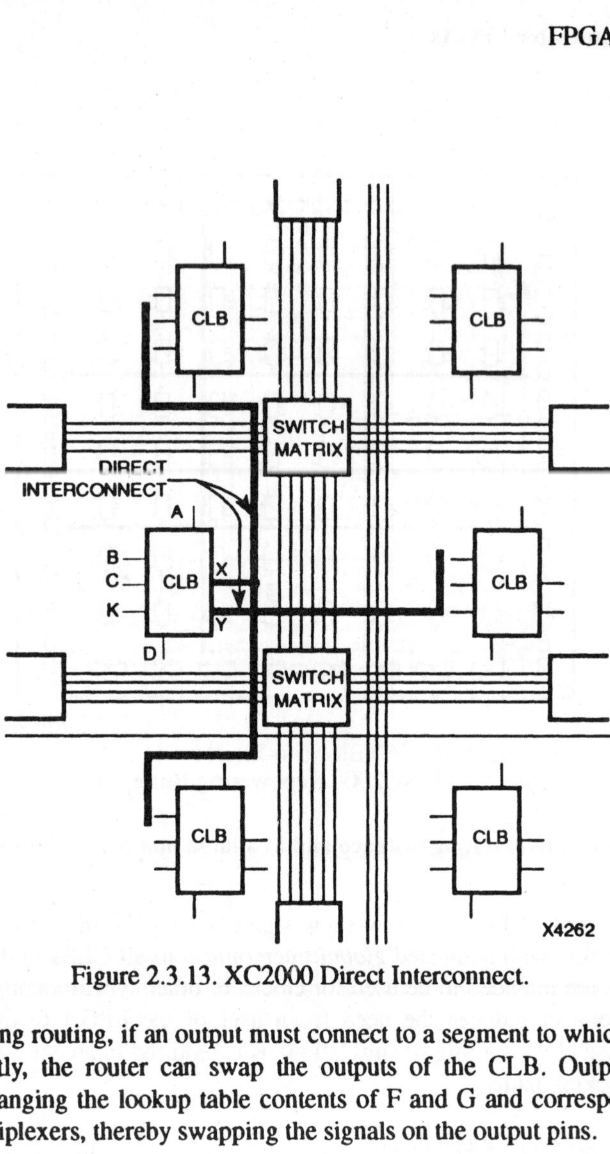

Figure 2.3.13. XC2000 Direct Interconnect.

problem. During routing, if an output must connect to a segment to which it does not connect directly, the router can swap the outputs of the CLB. Output swapping involves exchanging the lookup table contents of F and G and corresponding input selection multiplexers, thereby swapping the signals on the output pins.

XC2000 Family Members

The XC2000 family consists of two members, the XC2064 and the XC2018 (see table). The maximum gate capacity number is a rough estimate that assumes full utilization of all CLBs on the chip. No one design will use all the logic capacity of every CLB, so the typical gate capacity is lower than the maximum.

The XC2000 architecture has not changed since 1985, but improvements in the design and processing have improved the performance of the parts significantly over that period. XC2000 FPGAs operate at flip-flop toggle frequencies up to 100 MHz. The toggle frequency is the speed path from a CLB flip-flop, back through the lookup table and into the flip-flop. It is a measurable maximum frequency at which the part is

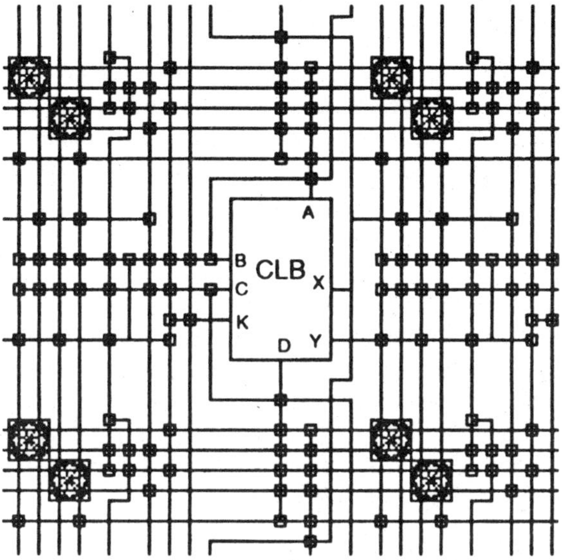

Figure 2.3.14. XC2000 Block-to-Interconnect Connections.

Table 1: XC2000 Family Members

Member	CLB Array Size	IOs	Gate Capacity	
			max	typical
XC2064	8x8	58	1200	800
XC2018	10x10	74	1800	1200

guaranteed to operate. More complex functions require inputs from other sources, and since those inputs must come through the interconnect, they will be slower. Typical worst-case system frequencies depend on the number of lookup tables and on the wiring in the critical speed path. Typical system clock speeds for the XC2000 are about one-third to one-fourth the maximum toggle frequency.

The Xilinx XC3000 Architecture

The XC3000 family of FPGAs is the most widely used FPGA family, with offerings from Xilinx, Seiko and AT&T. The XC3000 architecture includes enhancements to the XC2000 architecture to improve performance, density and usability. The result is a powerful collection of logic capability in a cost-effective package. Key to the popularity of the XC3000 family is the capacity of the parts, reaching well into the thousands of gates.

The XC3000 architecture was developed with manual tools for design implementation, and the architecture shows a bias toward manual design. This bias is exhibited primarily in the patterns of connectivity in the interconnect. These patterns are usable by a human designer, but provide a difficult set of constraints for software design automation.

The XC3000 CLB

The XC3000 CLB (figure 2.3.15) is substantially larger than the XC2000 CLB. Each of the lookup tables has four inputs rather than three, hence requires sixteen bits of configuration memory rather than eight. The two lookup tables can be combined with a multiplexer to produce any function of five inputs and some functions of up to seven inputs. The wider functions of the CLB extract only a slight delay penalty, and they allow the XC3000 architecture to implement faster logic, since speed critical paths can be implemented with fewer CLBs in series.

The XC3000 CLB has two flip-flops, to ensure that all combinational logic can be followed by a pipelining flip-flop. As in the XC2000, the flip-flop outputs can be fed back internally in the CLB to serve as inputs to the lookup tables. This register-rich

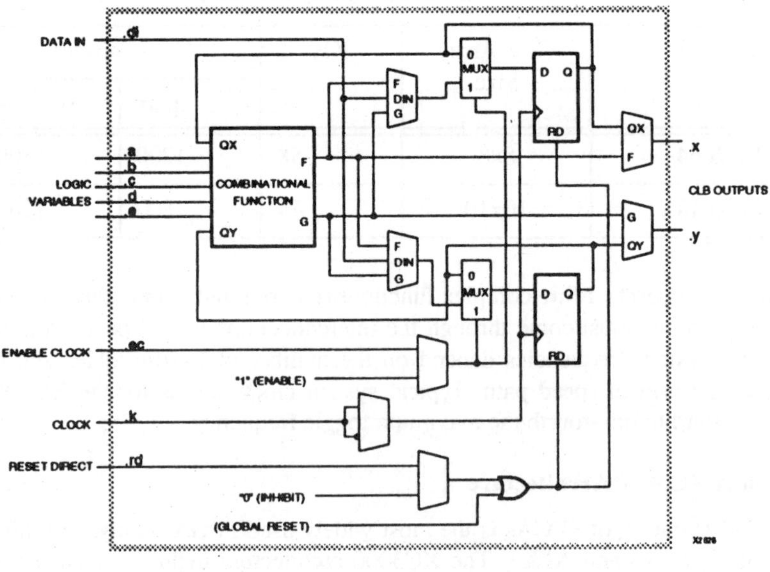

Figure 2.3.15. The XC3000 CLB.

cell lets the XC3000 implement state-intensive applications and heavily pipelined designs efficiently.

The XC3000 IO Block

The XC3000 IOB, shown in figure 2.3.16, is significantly more complex than the XC2000 IOB. The most important addition is a flip-flop in the output path. By registering data in the IOB, the clock-to-out time does not include interconnect delays. The result is a fast, predictable clocked output. The XC3000 IOB also includes a programmable pullup, optional output inversion and selectable slew rate. A slower I/O slew rate for low-speed signals reduces power surges, simplifying board-level design.

Inputs from the pad can be brought into the interior of the chip either directly or registered or both. By allowing both, the I/O block can de-multiplex external signals such as address/data busses, storing the address in the IO flip-flop and feeding the data directly into the wiring.

Wiring Architecture

The XC3000 interconnect structure shown in figure 2.3.17 has five general interconnect lines both vertically and horizontally. In addition, each CLB has direct connections to adjacent CLBs both vertically and horizontally.

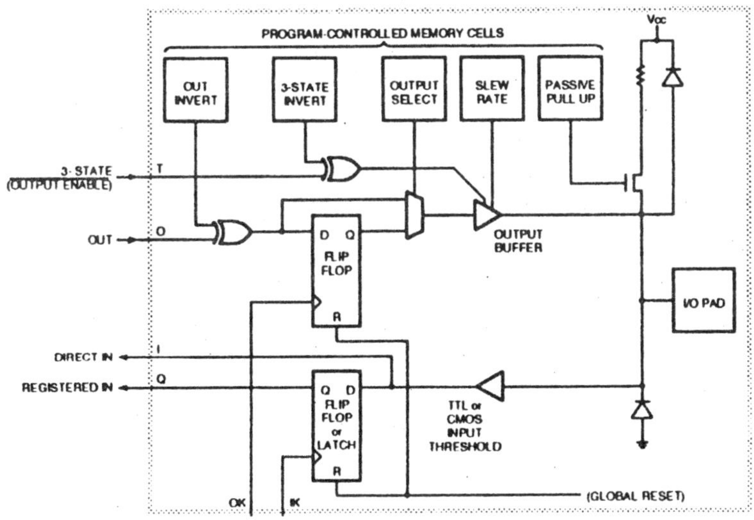

Figure 2.3.16. The XC3000 IOB.

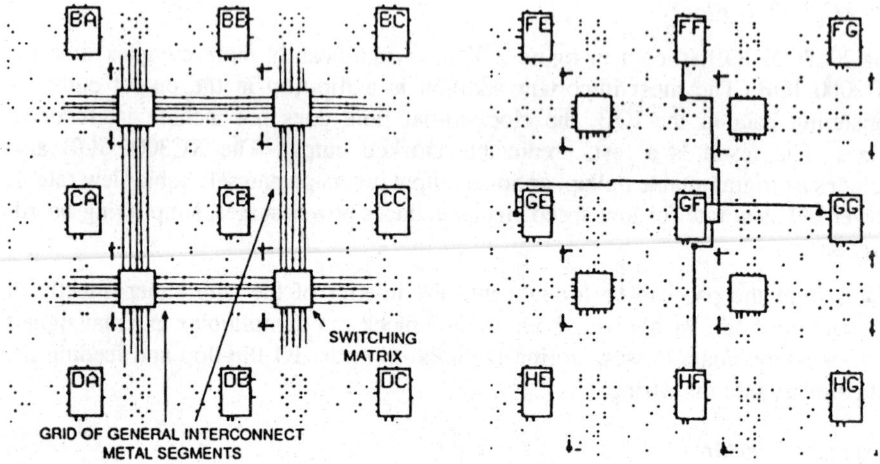

a

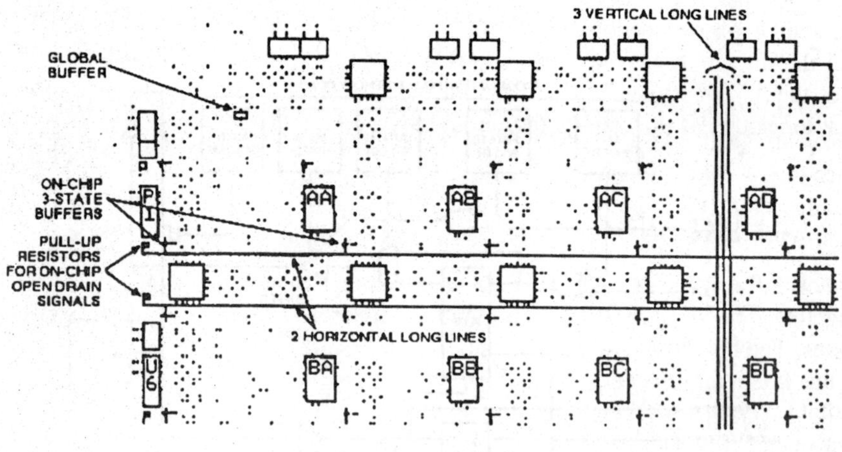

b

Figure 2.3.17. XC3000 Wiring.

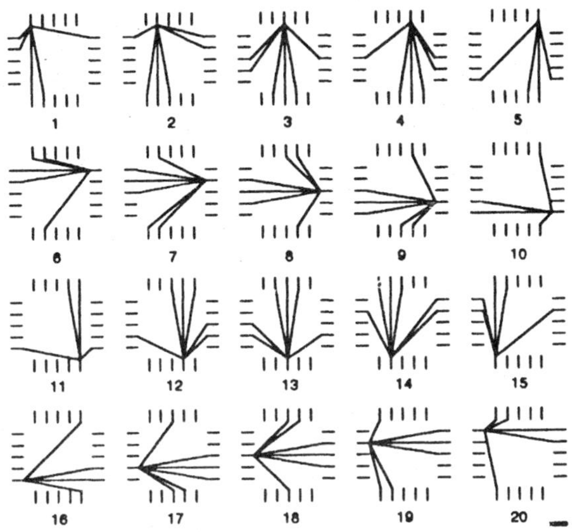

c

Figure 2.3.17 continued. XC3000 Wiring.

The switchbox connections in the XC3000 are more uniform across the segments, with the interconnect pattern including all wiring tracks. This additional flexibility gives the XC3000 router the ability to avoid blockages by moving signals from one track to another as they are routed across the chip.

Since the XC3000 tile was intended to be built into larger arrays than the XC2000, it includes more long interconnect. There are three vertical and two horizontal long lines. As in the XC2000, signals are always buffered when driving long lines.

Two of the horizontal long lines are driven by three-state buffers, distributed along the long line, one set per CLB. These buffers give users the ability to build on-chip busses. The busses allow implementation of datapaths without the wiring congestion of the wide multiplexers that would be required if three-state busses were not available. Each horizontal long line includes an optional pullup resistor to pull the long line high if none of the buffers is driving it. Configuring the three-state buffers open-drain, connecting the same signal to the input and control of the three-state buffer, and activating the pullup resistor creates a distributed wired-AND.

The three-state buffers on the long line are simple to implement because they are made up of programming features that are already present in the chip. All long line buffers are necessarily three state buffers, because only one signal drives the long line in standard use. In the XC2000 the three-state control is connected to a programming cell only. The XC3000 includes an optional alternate connection to the three-state driver that comes from the interconnect (figure 2.3.18).

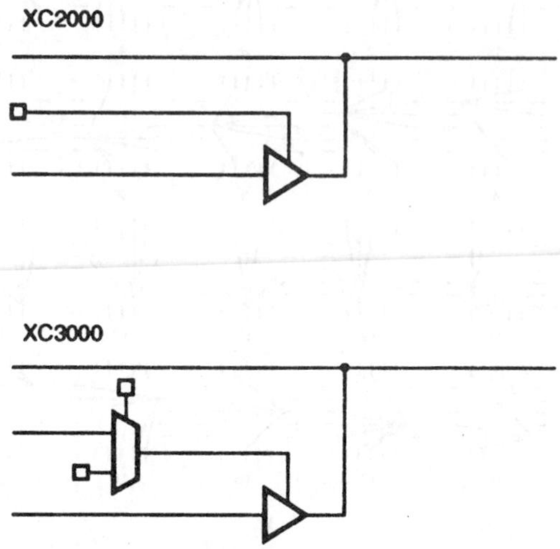

Figure 2.3.18. XC2000 Long Line Buffer and XC3000 Three-State Long Line Buffer Circuitry.

Experience with the XC2000 repowering buffers showed them to be difficult to use effectively. Signals should be buffered after they pass through several switchboxes, but the XC2000 architecture re-drives signals whenever they cross the grid boundary, regardless of the distance they have travelled. A signal to be connected to an adjacent CLB is buffered if it happens to fall at the edge of the grid pattern, whereas a signal that travels a long distance wholly within one of the squares of the grid will not be buffered at all. In both cases, the signal is slower than it would be with intelligent buffering. The XC3000 solves this problem with a combination of hardware and software. The XC3000 interconnect contains redrive buffers scattered among wiring segments. The redrive buffers can be configured to drive a signal from either direction along a general interconnect line. A timing-sensitive router selects buffered and unbuffered segments as appropriate to improve the performance of the net.

Block-to-Interconnect Connections

As shown in figure 2.3.19, the XC3000 architecture has a complex interconnection scheme from the wiring channels to the CLB inputs. Rather than group all connections in a single channel as in the XC2000, the inputs are connected to segments in two wiring channels, and each input pin connects to a fraction of the segments in the channel. A net routed in either channel can be connected to the input pin, provided it is on the proper track in the channel. Similarly, as shown in figure 2.3.20, outputs from CLBs can be connected to two different channels. The channels are at the ends of the fixed output wires shown in the figure. The output channels are not both adjacent to the CLB. This enlarges the immediate neighborhood for high speed connections between CLBs, since a signal can skip a switchbox in two of the four directions.

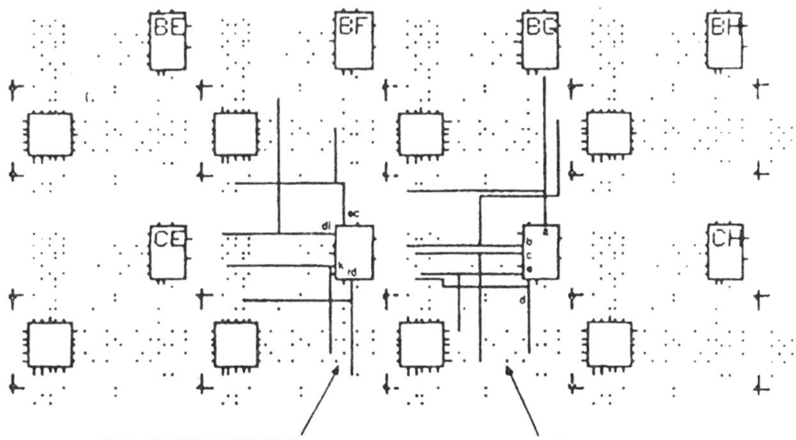

Figure 2.3.19. Input Pips in XC3000.

Both of these complexities cause placement problems. The placement software measures the quality of the placement using an estimate of routability and performance. For the XC3000, the pins are accessible from more than one channel, but there is no guarantee that the signal can be connected to the pin in any channel. Therefore, the routability depends on which channel the placer expects the router to use to route to the pin and on the ability of the router to bring the signal into the channel on the correct track.

The XC3000 Family Members

The XC3000 family consists of six members, covering more than a factor of six in capacity (see table 2). Devices are available in a variety of packages, including pin grid arrays, quad flat-packs and thin flat packs for PC-MCIA applications. Many members are available in common packages with identical footprints so designs can be migrated to higher or lower density parts without any board changes.

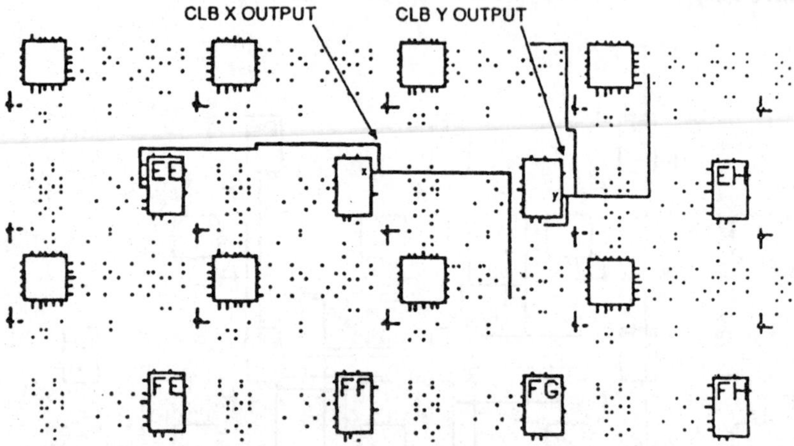

Figure 2.3.20. Output Pips in XC3000.

Table 2: XC3000 Family Members

Member	CLB Array Size	IOs	Gate Capacity	
			max	typical
XC3020	8x8	64	2 000	1 200
XC3030	10x10	80	3 000	1 800
XC3042	12x12	96	4 200	2 500
XC3064	16x14	120	6 400	3 800
XC3090	16x20	144	9 000	5 500
XC3195	22x22	168	13 000	7 500

Since its introduction in 1987, the speed of XC3000 FPGAs has increased with process migration from 1.2μm to 0.8μm feature size, and with circuit improvements to speed-critical paths. Performance of the XC3000 FPGAs over time is plotted in figure 2.3.21. Currently, the highest-speed parts are the compatible XC3100-3 (for 3ns CLB delay). This speed permits system clock frequencies around 100 MHz, although some designers have reported achieving speeds even faster than that on pipelined designs with careful partitioning, placement and routing.

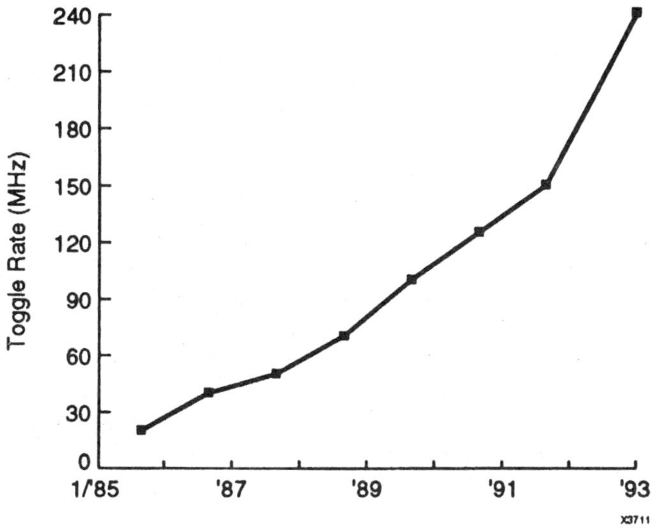

Figure 2.3.21. Historical Performance of XC3000 Parts.

Additional enhancements to the XC3000 speed and density came as a result of software improvements. These improvements do not change the maximum gate capacity, nor do they change the maximum toggle frequency, but they do increase the typical capacity and narrow the difference between the toggle frequency and the automatically-achievable system clock frequency. Software has improved the speed of automatically placed and routed designs by about 50% since the introduction of the family.

The Xilinx XC4000 Architecture

The XC4000 architecture [Hsieh 1990] [Trimberger 1991] was designed to improve performance and gate density for large designs. Several dedicated features were added to the general-purpose logic features of the XC3000, yielding an interesting combination of special-purpose and general-purpose functions. The XC4000 family was designed using placement and routing tools to evaluate architectural decisions.

Architectural features were designed to interact efficiently with an automated design methodology.

The XC4000 CLB

The XC4000 CLB (figure 2.3.22) is similar to the XC3000 CLB. It contains three lookup tables and two flip-flops. The two primary lookup tables, labelled F and G, each implement any function of four variables. These two results can be brought out of the block independently or they can be combined with another input in the H lookup table to make any function of five inputs or some functions of up to nine inputs. This allows functions such as nine-input AND, OR, exclusive-OR (parity) or address decode to be done at high speed in one block. The flip-flops can take their inputs independently from the lookup tables or from external signals, but they share control signals. Unlike the XC2000 and XC3000, flip-flop outputs are not recirculated internally. A registered feedback signal in the XC4000 must be routed in the general interconnect back to a CLB input pin.

The XC3000 can implement arithmetic with sum in one lookup table and carry in the other. The XC4000 CLB can implement arithmetic this way also, but since the speed of arithmetic operations is dominated by the speed of the carry chain, the XC4000 CLB includes dedicated high-speed carry logic. Figure 2.3.23a shows a configuration of the CLB in arithmetic mode. The F and G lookup tables compute two sums while dedicated logic calculates the carries. The dedicated carry logic in the XC4000 substantially speeds up arithmetic while doubling its density.

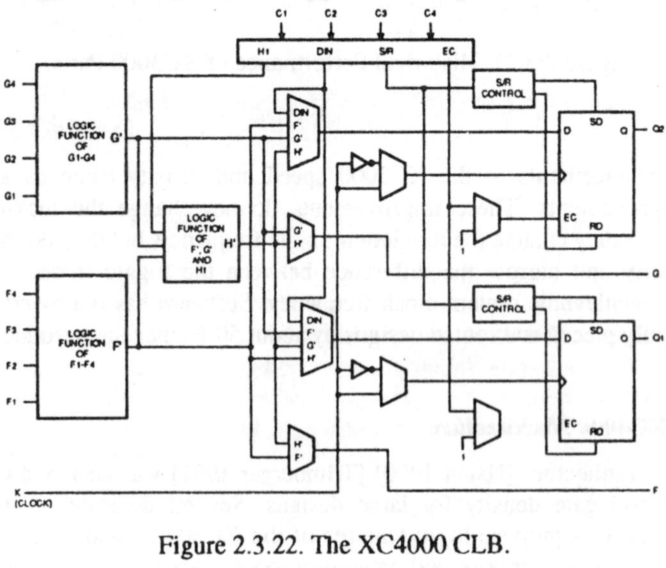

Figure 2.3.22. The XC4000 CLB.

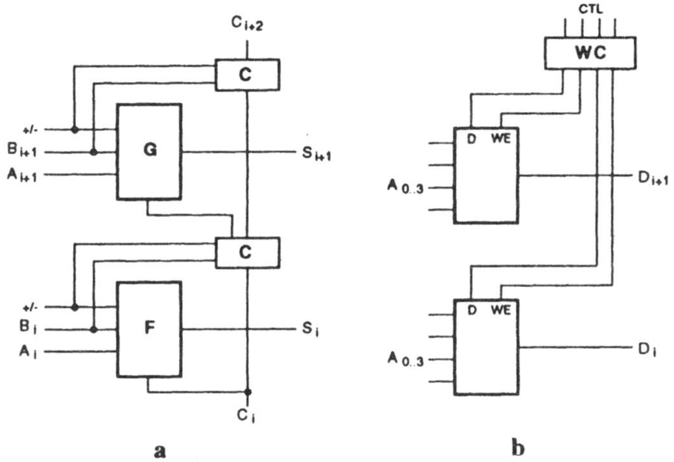

Figure 2.3.23. Special Configurations of the XC4000 CLB. a) Arithmetic. b) Memory.

Small, fast memory is an important component of many digital systems, appearing frequently as registers and FIFOs. The F and G lookup tables in the XC4000 CLB can be configured as 16x1 static memories. To write into the lookup table memory, the lookup table inputs address the memory cell to be written. Additional inputs to the CLB are write enable (WE) and data (D) (figure 2.3.23b). Reading from a memory is the same as evaluating a function. One CLB can implement 32 bits of memory, configured either as 16x2 or 32x1. On-chip memory makes FPGAs a reasonable implementation target for register-intensive digital systems. Since the memories are traded off against logic, unused memory is not wasted.

The XC4000 IO Block

Figure 2.3.24 shows the details of the XC4000 IOB. Signals to be output from the chip can be registered before output and enabled by a separate control signal. Outputs can be optionally pulled up or down, and the output driver can be configured with either fast or slow slew-rate. Inputs from the pad can be brought into the interior of the chip directly, registered or both to facilitate multiplexed bus interfaces. Furthermore, inputs can drive dedicated decoders, built into the edge interconnect, for fast recognition of addresses.

The XC4000 IOB includes boundary scan logic compatible with the ANSI IEEE 1149.1 (JTAG) boundary scan standard [IEEE 1989] [Maunder 1990]. The logic is not shown in figure 2.3.24. Boundary scan can check internal logic or external logic. Scan operations can take place before or after the FPGA is programmed and do not interfere with the operation of the part. The scan path and control signals are available

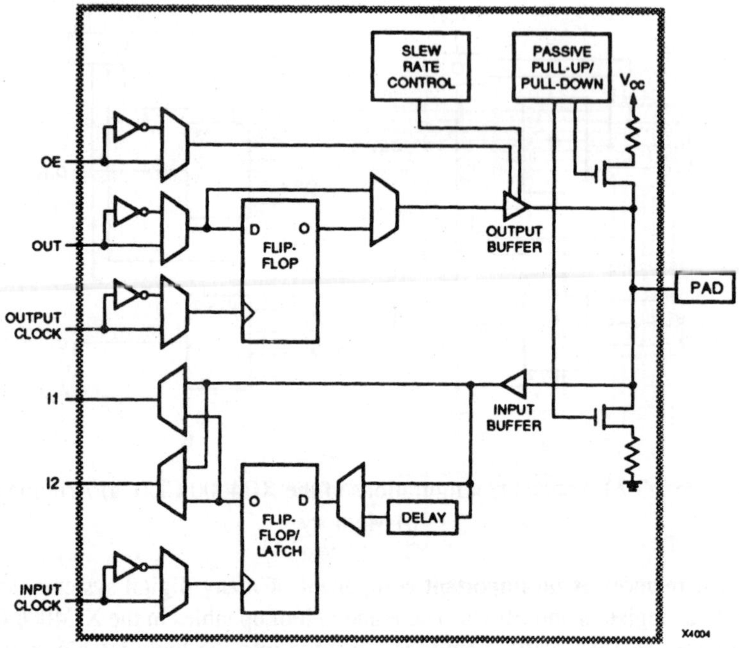

Figure 2.3.24. The XC4000 IOB.

internally in the FPGA, so additional test paths and registers can be implemented on the FPGA to make customized tests.

Wiring Architecture

Figure 2.3.25 shows an overview of the general-purpose interconnect in the XC4000. This wiring includes single-length *general-purpose* interconnect, like the XC2000 and XC3000, and *double-length lines* that bypass alternate switchboxes. Since signal delay is more dependent on the number of pips through which a signal passes than on the length of the segments, the double-length lines allow a signal to travel twice the distance in the same amount of time, or to travel a given distance in half the time.

The switchbox connections in the XC4000 are significantly fewer than those in the XC2000 and XC3000. Inside the switchbox, each segment can connect to three others, one on each of the other three sides of the switchbox (figure 2.3.26). Fewer pips means less load and faster interconnect. The drawback of fewer pips is that wiring may be more difficult. As a result, the XC4000 has more interconnect segments in the channel than an equivalent XC3000 would have.

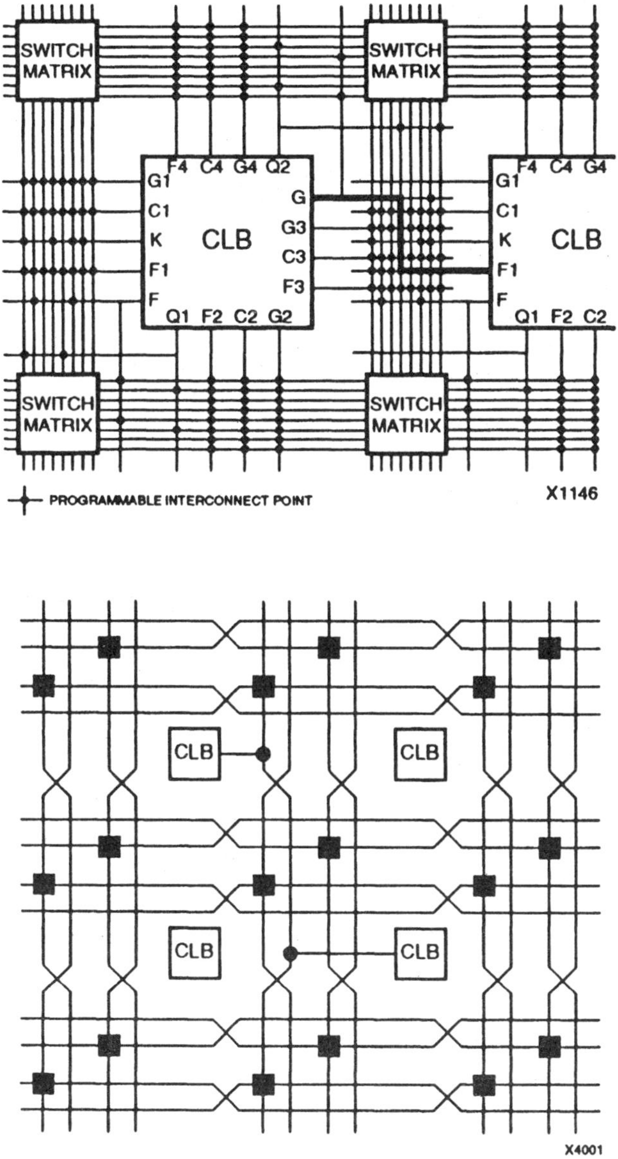

Figure 2.3.25. XC4000 Interconnect.

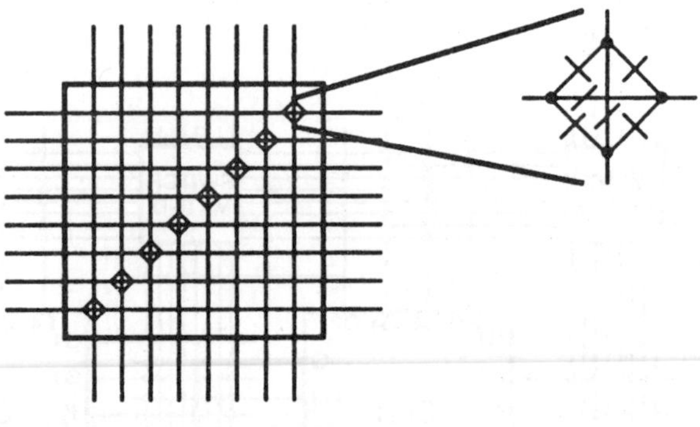

Figure 2.3.26. The XC4000 Switchbox Connections.

The XC4000 interconnect includes more long lines and global lines than were available in the XC2000 and XC3000 for high fanout and high-speed wiring. Two of the long lines in each row can be configured as three-state busses. The XC4000 long lines can be broken in the center of the chip to provide two half-long lines to improve routability. The half-long line has half the capacitance of the complete long line, thus decreasing the delay.

Four of the vertical long lines in figure 2.3.27 are the *global long lines*. Signals on global long lines can originate on-chip or off-chip. They are driven by dedicated high-

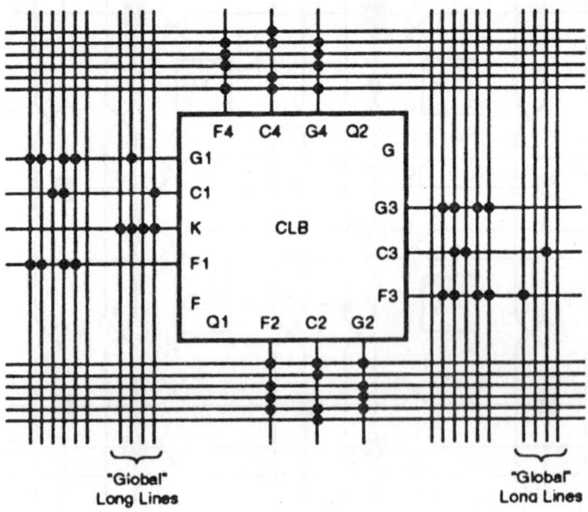

Figure 2.3.27. XC4000 Long Distance Interconnect.

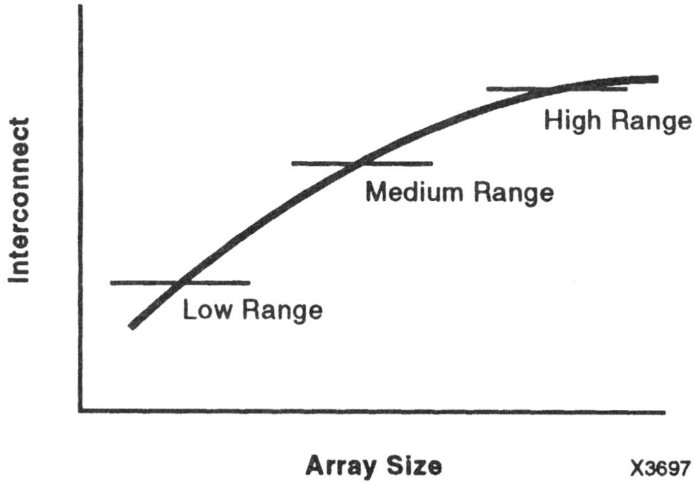

Figure 2.3.28. Interconnect Requirements Versus Array Size.

drive clock buffers and wired through the core of the chip to all CLBs to minimize clock skew.

Large arrays of blocks require proportionally more interconnect than small arrays. Therefore, the XC4000 family has three different size ranges of interconnect (see figure 2.3.28). The middle range, designed to support 14x14 to 22x22 arrays of CLBs, has eight single-length lines, four double lines and six long lines in each channel horizontally and vertically. It has four additional clock lines per column. The interconnect resources for the smaller range of parts has four single-length lines, four double lines and four long lines in each channel. The channel description for the larger parts has not been announced, but wirability estimates suggest that the differences may be significant. Using Heller's [1978] techniques, approximately 24 tracks are needed to successfully route a 25x25 array consistently.

Unlike its predecessors, the XC4000 has no redrive buffers in the interconnect. Instead, a signal requiring a repowering buffer must be routed through a CLB, either through an unused lookup table or through a repowering path that uses a CLB control input and sequential output. This design eliminates some of the complexity of the interconnect, at the cost of requiring timing-driven placement and routing software to make the trade-off between routability and performance of signals.

High-Level System Support

The array of CLBs naturally divides into two-bit slices for datapath operations (figure 2.3.29). The logic blocks contain carry logic to support two bits of addition or subtraction per block, or two 16x1 memories and two flip-flops per block. There are

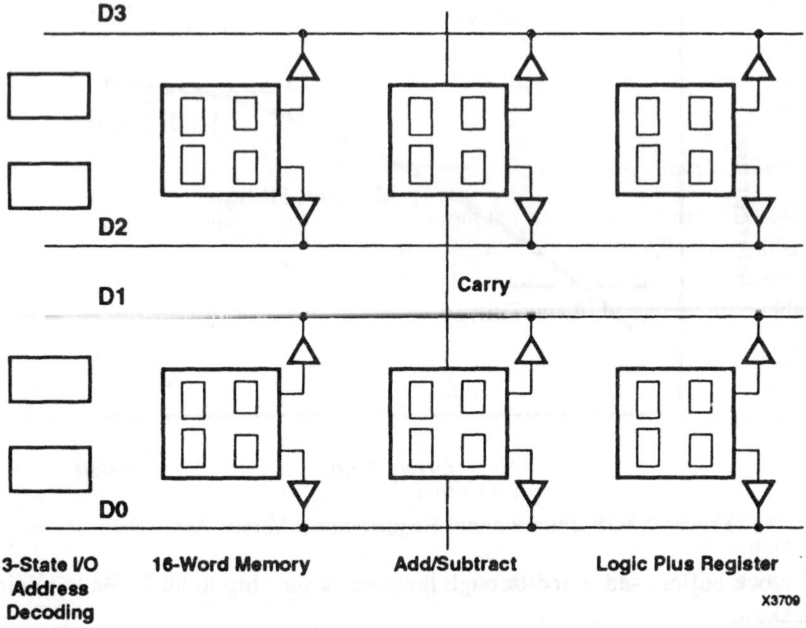

Figure 2.3.29. Bitslices in the XC4000.

two three-state bus lines per row in the array. These bus lines connect to two I/O blocks per row. I/O blocks can be configured to continue the three-state bus off-chip. Interfaces to external systems are simplified by bus-de-multiplexers, on-chip address decoders and low-skew on-chip clocking.

As a result, the XC4000 architecture conveniently supports bus-oriented datapaths, bit-sliced horizontally, two bits per row. Datapaths can contain arithmetic, memory or any other function implementable in a lookup table or a collection of lookup tables. This functionality is available in addition to the ability to build random logic, interconnected in whatever fashion is needed, using lookup tables and flip-flops.

Block-to-Interconnect Connections

The XC4000 block-to-interconnect connections are very similar to those in the XC2000. Block inputs connect to all interconnect segments in the channel, but outputs connect to only half the segments (figure 2.3.25). As in the XC3000, the CLB outputs drive signals in two different channels, but both those channels are adjacent to the block, simplifying the placement wirability estimation. The XC4000 inputs and outputs are distributed around all four sides of the CLB, eliminating the directionality bias of the XC2000 and XC3000.

The control signals to the XC4000 CLB are connected to wiring channels on all four sides. Configuration circuitry allows any of the control pins to drive any of the internal control signals, DI, CE, INIT or H1. The design software recognizes these control signals as logically equivalent, and uses pin swapping to improve delay and routability.

The XC4000 Family Members

The XC4000 family consists of ten members, covering a wide range of logic capacity (see table). Smaller parts (XC4002A through XC4005A) are available with less interconnect, and hence a smaller die. Two parts, XC4003H and XC4005H are available with increased I/O capacity.

The XC4000 CLB delay is 4.5 ns, comparable to an XC3000 toggle rate of about 180 MHz. In this architecture, systems in the 50 MHz to 60 MHz range are practical, and much faster systems are possible.

Table 3: XC4000 Family Members

Member	CLB Array Size	IOs	Gate Capacity	
			max	typical
XC4002A	8x8	64	3 000	2 000
XC4003A	10x10	80	4 500	3 000
XC4003H	10x10	160	4 500	3 000
XC4004A	12x12	96	6 000	4 000
XC4005/5A	14x14	112	7 500	5 000
XC4005H	14x14	192	7 500	5 000
XC4006	16x16	128	9 000	6 000
XC4008	18x18	144	12 000	8 000
XC4010	20x20	160	15 000	10 000
XC4013	24x24	192	20 000	13 000

Programming the FPGA

Configuration

Configuration is the process of loading the program into the FPGA. The program of an SRAM-programmable FPGA is stored in an array of memory cells that make up the lookup tables and control the pips and multiplexers. The program is loaded into the FPGA serially to minimize the number of pins required for configuration and to reduce the complexity of the interface to external memory. Internally, the configuration memory is arranged as a two-dimensional array. The program is shifted into the FPGA and assembled into a long word that is then loaded in parallel into the configuration memory. During programming, all internal drivers are disabled to avoid potential contention in the interconnect. Contention, even momentary contention, causes power surges, and the high current can damage internal metal traces.

All Xilinx FPGAs include the circuitry necessary to load the program from external memory. Essentially, every FPGA includes its programmer. The FPGAs can be configured in a number of different ways, depending on the application (figure 2.3.30). Configuration can be initiated by the chip itself when it senses power-up. In this "master" mode, the FPGA uses a simple 4-wire interface. The chip generates `clock` and `chip enable` signals to extract data from a small-footprint serial PROM which produces the program serially on the `data` line. In a system, the FPGA may be permanently mounted on the board, with the programming bitstream stored in the replaceable PROM. If the designer prefers to use an external clock to control programming, the chip can be configured in *slave* mode, where an external source generates the clock and data signals. The FPGA can emulate a simple processor peripheral for configuration, accepting an external clock, peripheral select and byte-wide data. In all modes, many chips may be connected in a daisy-chain fashion, with the first chip in the chain generating control signals and passing data to following chips. All configuration flows through the first chip and all chips become active simultaneously when programming is complete.

Besides configuring on power-up, SRAM FPGAs can be re-configured on command while residing in the circuit. This feature allows a designer to create a system in which the FPGA's program may change during operation.

Readback

Another powerful feature of the FPGA is *readback*, the ability to read out of the chip the program and also the contents of internal flip-flops, latches and memories. A working part can be stopped and its state recovered. In systems, readback is used for design verification and debugging. During manufacturing test, configuration and readback allow direct test access to observe and control internal nodes, greatly simplifying test program generation.

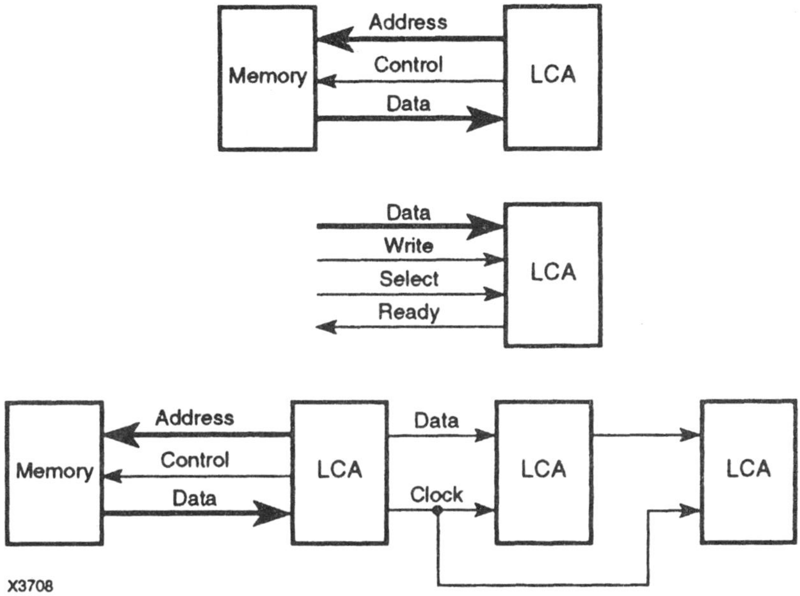

Figure 2.3.30. Configuration Patterns.

Security

The programming bitstream includes a security bit to prevent readback, so a design can be delivered without the ability to unload it from the part. In an application where secrecy is paramount, a user may configure the SRAM FPGA before shipping, and keep the part powered with a battery. The user may then remove the programming source and ship the system in which the FPGA resides. In low-power standby mode, a battery can hold the FPGA programming for years. This capability allows users the ultimate in security because there is no way to reverse-engineer the logic on the part. Programming makes no physical modification to the chip, so it cannot be recovered. If the chip is removed from the battery or power supply, the programming is erased.

2.4. Software

The design process for SRAM-based FPGAs is similar to that for MPGAs. The design process has three parts: design entry, implementation and verification (figure 2.4.1). Design entry tools used for MPGA design can also be used for FPGA design. These tools include schematic editors, state-machine generators, and synthesis tools. The interface to the FPGA design implementation system is a netlist plus implementation constraints, such as timing requirements and pad placement. Implementation tools partition the logic to fit into the FPGA, find a good placement for the logic and route signals between the logic blocks. The implementation system generates back-annotated netlists with routing delays for timing analysis and other verification steps. Like MPGAs, FPGA designs can be verified by simulation, and reprogrammable FPGAs make in-circuit verification an attractive alternative to extensive simulation.

Automated Design Implementation

Logic implementation on the Xilinx-style FPGAs consists of three steps, partitioning, placement and routing. This section describes the three steps and the techniques that have been used to address them.

Partitioning

The *partitioning* task, also called FPGA *mapping*, is the mapping of the logic represented by the incoming netlist into the physical primitives implemented on the chip. Typical MPGA design implementation flow does not include partitioning. Instead, MPGA vendors provide a *library* of gates for schematic entry and the corresponding hard-wired cells for the physical implementation of those gates. The incoming netlist must contain only cells from that library. The cells are mapped one-

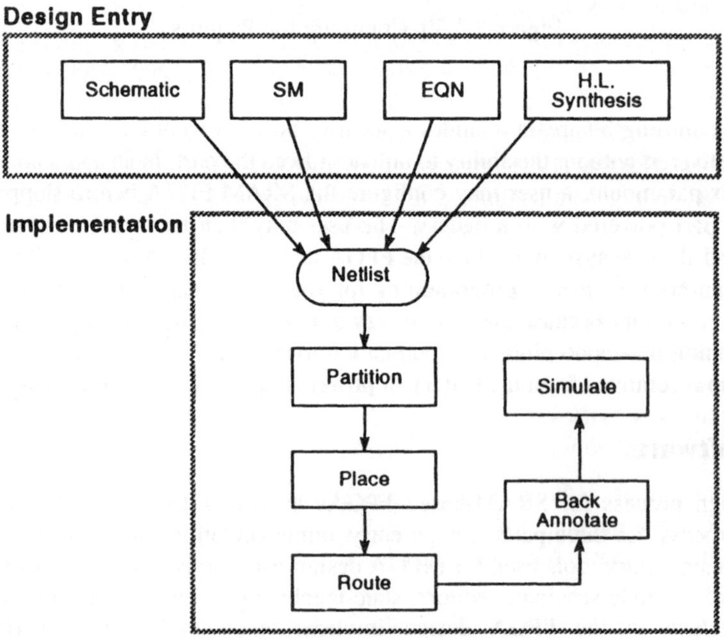

Figure 2.4.1. FPGA Design Flow.

to-one onto structures on the MPGA. While entering the logic design, the designer explicitly maps logic into the MPGA logic resources.

FPGA blocks are designed for logic density, so efficient manual mapping may be difficult. Each 4-input lookup table can represent 2^{16} functions. Each XC4000 CLB can represent approximately 10^{15} different functional patterns. Obviously, it is not reasonable to enumerate this library. Instead, the manufacturer supplies a smaller library of functions, called the "design library," and provides partitioning software to automatically map the logic in the design library into the structures on the FPGA that implement the logic. The Xilinx implementation system accepts a design in the form of a netlist of MPGA-like gates, and partitions the netlist, grouping gates into legal CLBs and IOBs for placement.

Many attempts to address this problem begin with technology mapping from logic optimization. Technology mapping is the translation of "technology-independent" logic, perhaps in the form of NAND gates, into an efficient set of gates in a target library. Several authors have reported work on mapping logic into lookup tables. Murgai [1990] uses Roth-Karp decomposition and kernel extraction in MIS-II to bound the number of inputs to nodes, then uses binate covering to group nodes into CLBs efficiently. Subsequent work [Murgai 1991] added additional covering techniques and cofactoring. Addressing performance of the resulting circuits, Murgai [1991b] decomposes the network and uses routing estimation and critical-path packing to reduce delay. Francis [1990, 1991a] decomposes the netlist into fanout-free trees that are then mapped using dynamic covering and bin packing. Additional improvements include checks for reconvergent fanout. Further work [Francis 1991b] optimized for performance by minimizing the depth of trees rather than number of lookup tables. Karplus [1991] transforms a netlist into an if-then-else directed acyclic graph (DAG), a generalization of a binary decision diagram. The DAG is then mapped using a simple marking process to achieve good results. Woo [1991] partitions a design into pieces of manageable size, then exhaustively enumerates cases to find which nets should be subsumed into lookup tables. Sawkar [1992] casts the problem as clique partitioning of the netlist. Cong [1991] [1992] decomposes the network into two-input gates and assigns a topological level (depth) starting at the primary inputs, increasing the level when the number of inputs exceeds the size of the lookup table. A backward mapping then constructs lookup tables, potentially duplicating logic to minimize network depth.

All these techniques address mapping logic into lookup tables, but do not handle the larger problem of mapping logic and flip-flops into CLBs. This problem is interesting because there is both a logical and physical component to the partitioning. The connections within a CLB are somewhat constrained, but the quality of the resulting partitioning depends on the quality of the subsequent placement, so physically-related logic should be partitioned together.

Xilinx uses two partitioning algorithms, targeted to different families of parts. The different algorithms result from the different types of intra-block constraints in the FPGA architectures.

Xnfmap is a partitioner targeted to the Xilinx XC2000 and XC3000 families. Xnfmap is driven primarily by logical constraints -- finding groups of logic that fit into a single CLB. Secondarily, Xnfmap attempts to maximize the number of nets collapsed into lookup tables and CLBs. Since collapsed nets require no interconnect, the more there are, the easier the task of routing the remainder.

Xnfmap distinguishes between combinational logic gates, which will be mapped into lookup tables, and all other logic elements, which are implemented directly by resources on the FPGA. These other logic elements include flip-flops, I/O pads and three-state buffers. Xnfmap builds groups of combinational logic by tracing back from the nets on the inputs of non-combinational-logic gates. Xnfmap adds the gate that sources the net, and recursively groups the inputs of the gate it found. The recursion stops if the net is not sourced by a combinational logic gate or if it fans-out to multiple destinations and therefore must be routed in the interconnect between CLBs. This is the case for the output of gate A in figure 2.4.2. In this way, Xnfmap groups all combinational logic into "cones" of logic that can be implemented in a lookup table.

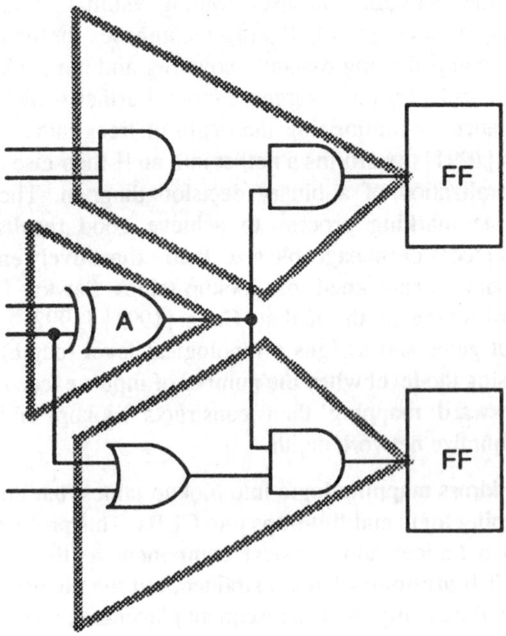

Figure 2.4.2. Cones of Combinational Logic Feeding Flip Flops.

After constructing logic groups, Xnfmap breaks up groups that are too big to fit into a single lookup table. It examines every possible division of the combinational logic to find the one that collapses the most nets, minimizes the number of groups or maximizes the use of CLB internal feedback paths. A single group may be broken several times before all of its pieces are small enough to fit into CLBs. At this point, each flip-flop is bound to the group of combinational logic that drives it.

Xnfmap then examines every possible legal pairing of flip-flops and logic blocks to find those that can be merged into a single CLB. Legal pairings are determined by the CLB architecture and are constrained by the number of outputs, the number of inputs, the number of flip-flops and the pattern of connections among them. Merging of legal pairs is *greedy*, taking those pairs that share the most inputs, thereby minimizing the number of connections for the router. Ties are broken by logical locality.

The XC4000 CLB has weaker coupling among CLB elements than the XC2000 and XC3000 architectures. Each XC4000 lookup table can be wired independently, and the outputs of the combinational and sequential elements can be routed out separately. This reduction in constraints means the XC4000 partitioner, *Blkmake*, can partition based primarily upon routability criteria rather upon than the architecture of the CLB [Trimberger and Chene 1992].

Blkmake has four steps: mapping combinational logic into lookup tables, grouping pieces of logic with pattern-matching, generating a placement-driven preferred partitioning of lookup tables and other logic, and completing the partitioning based on interconnect constraints inside the CLB. This combination of physical and logical partitioning makes a high-quality, legal partitioning.

Mapping combinational logic into lookup tables is done with a modification of Francis's Chortle technology mapper [Francis 1990]. Chortle breaks combinational logic into fanout-free trees. Since a lookup table has only a single output, the output of a fanout-free tree must be a lookup-table output. The Chortle mapper scans from logic inputs to the output using a dynamic program with variables of number of lookup tables and number of lookup table inputs used. Each step in the dynamic program finds the minimum number of lookup tables required to map the logic gate using different numbers of inputs to the lookup table that generates the gate's output. The chosen result is the one that requires the fewest number of lookup tables.

Blkmake's enhanced version of the Chortle algorithm uses support-based mapping for greater compression of reconvergent-fanout logic. It identifies lookup tables that meet the constraints for the H lookup table and preferentially accepts them during the dynamic program, increasing the number of lookup tables, but decreasing the number of CLBs required to implement those lookup tables and decreasing the delay of combinational logic.

Finally, Blkmake selectively duplicates logic to eliminate fanout nodes, when the duplicated logic improves density and performance. Gate A in figure 2.4.3a may be

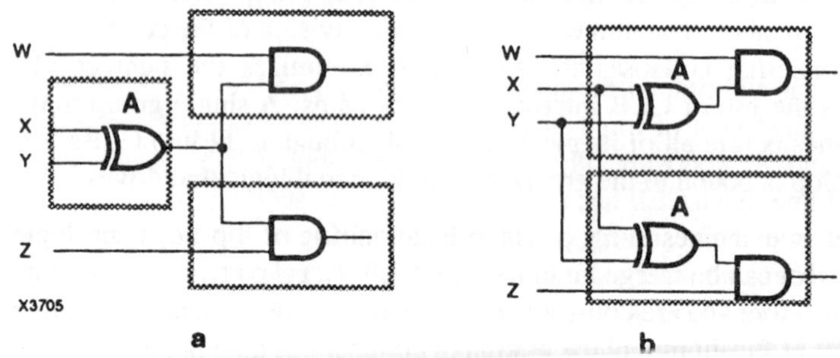

Figure 2.4.3. Duplicated Logic to Improve Performance and Density. a) Original Logic with Fanout. b) Duplicated Logic.

duplicated as shown in figure 2.4.3b, since the logic can then be implemented in two lookup tables rather than three. This logic duplication also decreases the path delay from two lookup tables with a connection between them to one lookup table with no interconnect.

Following lookup-table-mapping is a pattern-matching step to group logic that matches particularly-constrained structures in the CLB. Figure 2.4.4 shows a few of the groups that are built by pattern matching. Each group of logic is treated as a unit in subsequent placement-based optimization. The final step of pattern-matching groups lookup tables that share inputs. The grouping is greedy, grouping lookup tables with maximal sharing first.

At this point, the netlist consists of *block elements*: lookup tables, flip-flops, three-state buffers, I/O pads, clock buffers and other logical elements that correspond to physical resources on the FPGA. Some of these block elements may be grouped. The third step of partitioning uses mincut [Breuer 1977] [Fiduccia 1982] to generate a preferred placement for the block elements. Blkmake's mincut uses simple terminal propagation [Dunlop 1985] [Hartoog 1986] and a derivative of Sechen's crossing-

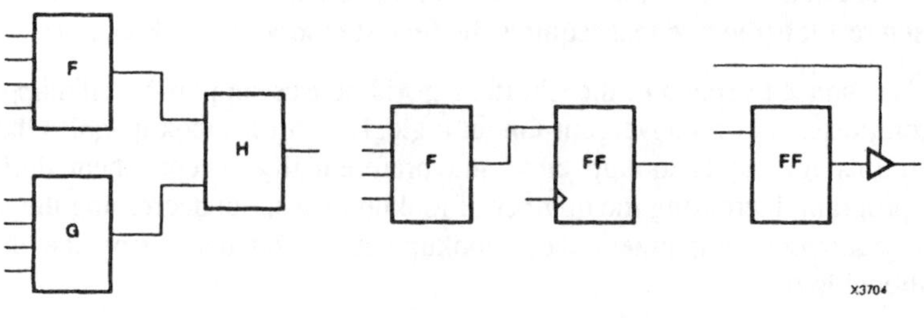

Figure 2.4.4. Pattern Matching Groups.

based cost function [Sechen 1988]. Mincut preserves the groups built by pattern-matching.

An important addition to the standard mincut algorithm is the preservation of the horizontal alignment of three-state buffer nets. The three-state buffers in the Xilinx XC3000 and XC4000 are aligned horizontally on the chip and drive a common horizontal line. Each CLB has two associated three-state buffers on two different nets. To ensure that the resulting CLBs contain matchable pairs of three-state buffers, Blkmake groups all three-state buffers that drive a common net, and treats the group as a single wide block. When the three-state buffer block no longer fits into a partition, it is locked into place. Blkmake makes a single, wide partition for the three-state bus line that contains only the three-state bus drivers and any logic that has been grouped with them. It continues to divide these new partitions, but only with vertical cut lines. Blkmake continues to partition the non-three-state logic with both vertical and horizontal cut lines.

Blkmake's mincut partitions the logic with some knowledge of the underlying FPGA architecture. It may divide the logic into uneven partitions, but maintains and manages constraints on the resources required by each of the partitions and moves the dividing line between partitions accordingly. Blkmake's mincut terminates when every partition represents a single CLB at a specific location on the target chip. However, it may not be possible to combine all the elements in the partition into a single legal CLB.

The final partitioning step is the construction of a legal CLB for the block elements in each partition. The construction is a multiple-pass greedy method that pushes overflows into adjoining CLBs. The construction is run many times from different random starting points, keeping the best result. The constructed CLB is placed at the location at which mincut placed the original partition.

Placement

The placement step accepts the partitioned design in terms of CLBs and IOBs and determines a good placement for the blocks in the FPGA array. Many MPGA placement algorithms are applicable to FPGAs with minor modifications. Most MPGA placers optimize with respect to a cost function that is based on the total expected length of all nets in the design. Net length is a reasonable measure because a shorter net length implies less interconnect area for interconnect and less capacitive load on nets. The result is that total net length correlates well with both routability and performance. However, in an architecture where the wiring can come in a few fixed lengths, as it does on all three Xilinx architectures, net length does not necessarily correlate well with either routability or performance.

Placement software recognizes two classes of constraints on placement due to the FPGA architecture: *legality* constraints which must be met to build proper functionality, and *quality* constraints that are preferred to take advantage of FPGA features. Legality constraints are enforced in the move set. Quality constraints are handled in the cost function.

Legality constraints in the XC3000 and XC4000 dictate that three-state buffers driving the same bus must be horizontally aligned. If they are not, the three-state buffer outputs cannot drive the same net, so the placement is unroutable. In contrast, a quality constraint states that high-fanout and low-skew nets should be routed on long lines. The buffered long line delivers a signal with lower resource cost and lower delay than a solution with shorter total wire length that uses general-purpose interconnect. The placer attempts to align instances on these nets vertically or horizontally. If the instances cannot be aligned, the design is still routable, so the placer does not rigidly enforce this constraint.

Simulated annealing [Kirkpatrick 1983] has been used in many MPGA placement systems ([Sechen 1985 and 1988] and others) and is applicable to FPGA placement as well. Although simulated annealing is slow compared to constructive methods, it produces a good result regardless of the initial placement, and produces a good solution with complex, varying and possibly-conflicting constraints. The simulated annealing cost function for FPGAs contains two components for different kinds of constraints, wire length and alignment. The wire length component reduces the overall routing resource requirement, alignment allows efficient use of long lines. During placement, high-fanout nets are scored using a request/grant mechanism for long lines. The annealing cost function bases the cost for a net on whether or not it expects the net to be routed on a long line. As placement progresses, some nets may become misaligned and others more aligned, changing the placer's routing expectation which changes the way the placer calculates the cost of the nets. When placement finishes, the placer passes its routing expectation to the router as a guide to ensure agreement between the two pieces of software.

Placement for the XC2000 and XC3000 families addresses the asymmetry of the blocks by measuring wire length from the channels into which the pins connect, rather than from the blocks themselves. For those pins that connect into multiple channels, a single representative channel is used as the basis of wire length and alignment estimation.

As described earlier, Blkmake generates an initial placement for partitioned logic for the XC4000. This initial placement is improved using a Generalized Force-Directed Relaxation (GFDR) algorithm [Goto 1981]. As with simulated annealing, the GFDR cost function prefers aligned nets, boosting the effectiveness of long and double-length lines. Since moves are rejected much less often, GFDR is much faster than simulated annealing. Therefore it is run exhaustively, iterating until no improvement is found. The placement results are of comparable quality to simulated annealing, but are achieved with much less execution time.

Routing

Traditional MPGA global routing and channel routing algorithms do not work well with the FPGA architectures described in this chapter. Global routers assume that the routing problem can be decomposed spatially so that after global routing, the detailed routing of each channel can be done without interfering with other channels. This model does not work with FPGAs because the limited connections in the switchbox constrain the assignment of segments in one channel based on their assignment in other channels. For example, in figure 2.4.5, a snapshot of the XC4000 interconnect, if a net is routed on the bottom track in the lower channel, then because of the limited connections in the switch box, it can only be routed on the left-most tracks in the right and left channels and the bottom track in the upper channel. A global router that ignores these constraints will not correctly decompose the routing problem for a detailed router, and the detailed route will fail.

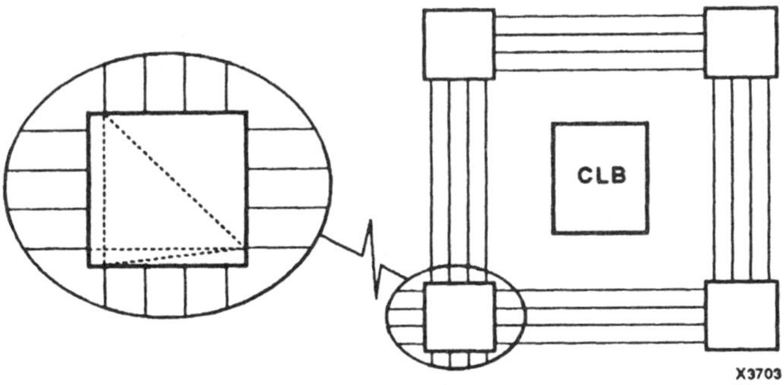

Figure 2.4.5. Routing Constraints Imposed by the Switchbox.

Channel routers assume that two non-interfering wiring layers are available for routing and that a net in a channel can change tracks when the route is infeasible due to cyclic vertical constraints [Soukup 1981]. It is possible in the architecture to provide a model of as many non-interfering wiring layers as needed. It is also possible to construct channels in which there cannot be cyclic vertical constraints. However, the limited connections in the switchbox and the limited connections from the CLBs to the interconnect channels prevent channels from being routed independently, and since the entire problem cannot be described as a single channel, channel routing is not applicable.

As a result, routing is done with an A* maze router on a graph [Nilsson 1971]. Graph nodes represent segments, and arcs represent pips and paths through CLBs. Routing proceeds from logic outputs to inputs. When searching for an input, any input to a lookup table is accepted and input pin swapping is performed. For combinational logic outputs, the router starts the maze search from both lookup tables in the CLB if they are available, to perform output pin-swapping. This ability to swap logically-equivalent pins significantly improves the routability of designs.

The router has a detailed timing model of the FPGA, including block delays, wire segment capacitances and pip resistances. The router keeps this information in an interconnect delay model based on Penfield and Rubenstein's work [1981]. It uses an incremental method of recomputing delays during wavefront expansion to measure and to reduce delay and skew while routing. This timing-driven feature is used in the XC3000 and XC4000 to determine when to route a signal through a redrive buffer or through the repowering path in a CLB.

The FPGA router includes a rip-up and re-route pass that is used both to complete routing and to reduce delay and skew on routed nets. After routing, the implementation system may re-place and re-route the design if the routed design does not meet all timing requirements.

Path delays, not net delays, dictate the performance of circuits. Frankle [1992] described the *limit-bumping algorithm* (*LBA*), a timing-optimization method using slack allocation among source-to-load connections in the design. Required path delays are specified between flip-flops, from I/Os to flip-flops, from flip-flops to I/Os and between I/Os. Required signal times are propagated back from path destinations to net destinations. Actual signal times are propagated forward from the path sources to each net destination. The difference between required time and actual time for a connection is called the *slack* of that connection. A single connection may be on several paths, so its slack is the minimum slack calculated over all paths. These propagations can be done in $O(c)$ time, where c is the number of connections in the design.

The LBA distributes a fraction of the slack on a connection to that connection. The fraction is pre-calculated as the ratio of the connection weight to the maximum weight of all connections on any path through the connection. The weight of a connection can be simply one, or it can be a function of netlist parameters such as net fanout. As long as the calculation of the fraction is based on the netlist it need not be recalculated during routing.

The LBA uses minimum delays for each connection as an initial lower bound and calculates slacks for all paths. It iterates using the fractional distribution algorithm above to allocate the slacks on paths to allowed connection delays until all slacks are near zero. At that point, the timing-driven router routes all connections to meet their delay constraints.

If delay constraints are not met by the router, the allowable delays are increased, and slacks allocated based on the new slacks introduced by the adjustment. If timing constraints are met, and we wish to produce a faster design, timing constraints are tightened using the same iterative method. The negative slacks are allocated to connections using an "approach fraction", which is proportional to the difference between the routed delay and the lower bound. The assumption is that a connection delay that is close to its lower bound cannot improve much, and so is not required to absorb as much of the negative slack.

The LBA has been applied to FPGA routing, with the measurement that it improves system performance by about 15% compared to connection-based timing-driven routing, and is within about 15% of optimum.

Technology-Specific Synthesis

The structured architecture of FPGAs implies preferred methods of implementing functions, particularly multi-bit functions. A schematic that is designed with the FPGA architecture in mind can be twice the density and speed as one that was designed without regard to the architecture.

Logic synthesis techniques address re-mapping random logic to a new architecture, but they do not address larger functions, such as memories and data path elements. A separate tool, called X-BLOX performs this function.

X-BLOX accepts a design as a netlist which includes variable-width multi-bit primitives. Users can specify data types: the width of busses and their encoding (such as two's complement or unsigned binary). X-BLOX determines the sizes of busses between primitives and the sizes of the primitives themselves if their widths and encodings are not specified, and checks compatibility of data types. It eliminates redundant logic and generates macros of the proper size for each of the operations. The macro generation is both technology-dependent and die-dependent. Different architectures prefer different implementations of the functions and different parts within a family have different numbers of clustered resources (such as column-aligned carry logic in the XC4000). X-BLOX generates its output accordingly, making optimized macros for each family member.

X-BLOX includes several other optimization steps related to the FPGA architecture. It moves flip flops from CLBs into IOBs where possible, assigns nets to global high-speed buffers if the device has them available.

A design using X-BLOX is simpler and faster to specify than one specified with a netlist of simple gates, because the desired functionality can be expressed in higher-level symbols. This high-level specification allows X-BLOX to optimize the high-level function to the target FPGA, so efficient design does not incur a density or performance penalty. X-BLOX can be used directly in a schematic or driven from high-level synthesis through a netlist.

Manual Design

MPGA and FPGA design systems allow manual intervention in the implementation to varying degrees. Typically, a designer can constrain placement to force signals to package pins. Additionally, a system designer can label nets as time-critical or assign nets and paths with maximum delay values. These constraints can be provided as annotation in the schematic or in a separate design constraints file.

Manual intervention in FPGA partitioning can take the form of a CLBMAP, a schematic-level constraint that forces the partitioner to accept a user-defined mapping for part of the logic. The CLBMAP is a cell in the library that a designer connects in parallel with the logic it is mapping. The CLBMAP represents no logic, but its connections are used to guide the partitioner to implement the mapped logic in a single CLB. A designer can go further, designing in terms of CLBs with their programming.

The XACT Design Editor (XDE) (figure 2.4.6) is an interactive graphical editor similar in concept to MPGA wiring editors. XDE contains both the logical and physical descriptions of the design. Users modify both descriptions simultaneously as they design the circuit. A designer can use XDE to pre-place and route CLBs or to do post-placement and routing fixup. XDE can also be used as a complete design system, allowing a designer to map the logic manually onto the device. XDE can accept a netlist as input, or a design can be created and implemented completely in XDE.

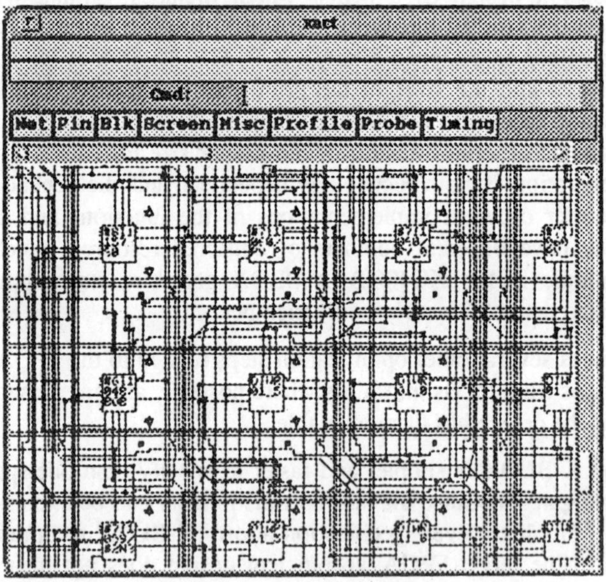

Figure 2.4.6. XDE Screen.

XDE includes the checking and editing functions required for manual design, including a design rules checker, a timing verifier and a router. XDE allows a designer to turn on or off individual pips in the interconnect, to set the functionality of function tables, and to control the functions in CLBs and IOBs.

XDE provides the front-end interface to the debugging system as well, allowing simple modifications of the design. It includes commands to generate *probe points*, internal signals routed out to unused pads to allow external access to internal nodes for debugging a design. The probe modifications are incorporated into a debugging version of the design that is then loaded into the FPGA for testing. XDE stores the probe points separately from the base design, so they can be eliminated easily when prototype debugging is complete.

2.5. The Future

FPGAs are similar to MPGAs in many respects, so there is a large body of knowledge that a researcher can draw upon to apply in this area. However, the space of FPGA architecture is large and relatively unexplored, providing many profitable areas for research. These unexplored areas may eventually yield significantly better FPGAs in terms of density and performance. Since software depends on the architecture, many software questions will opened or re-opened by FPGA architecture innovation.

Programming Technology

The CPLD-style array architectures, built with EPROM or EEPROM transistors, cannot be scaled beyond thousands of gates of logic. The array of transistors scales quadratically, as does static power consumption, while delays increase. All large-capacity EPROM devices show significantly degraded speed relative to smaller devices, as well as massive power requirements. Managing the size and power consumption requires a multi-level logic organization, such as the island-style architectures described in this chapter. EPROM transistors are only efficient when built in large arrays, so they become inefficient in these architectures. Recent attempts to extend EPROM-based architectures to large devices have separated the EPROM section into a straightforward memory array, and placed it next to an SRAM-based FPGA, basically building an SRAM FPGA with a monolithic PROM.

Large antifuse-programmed devices rely on very high reliability of the antifuses themselves. A single ten-thousand-gate antifuse-based FPGA may have over a million antifuses. Although only a few percent will actually be used by a design, the architecture relies on those few percent being correct. If they are not correct, the FPGA will fail to program correctly, and must be discarded. Discarding devices that fail to program is not a serious issue with small devices, where the parts cost ten dollars and programming yield is above 99 percent; but on large devices, the parts cost hundreds of dollars and the programming yield may be 80 percent. It is unlikely that customers will accept discarding 20% of their $500 FPGAs. The quality of antifuse manufacture limits the size of antifuse-based devices.

SRAM-programmed devices have none of these drawbacks. They scale well with technology improvements and have very low power consumption. They can be built with very high quality and fully tested at the factory. For these reasons, the capacity of

SRAM-based FPGAs will exceed the capacity of FPGAs based on other programming technologies.

Architecture

The FPGA architectures described in this chapter were developed in an attempt to balance density, performance, cost and ease-of-use goals. If one were to emphasize one or more of these goals at the expense of the others, a significantly better architecture might result. For example, wider lookup tables would allow faster functions of more variables, but increase the size of the chip, decreasing density and increasing cost. Narrower lookup tables reduce lookup table area, but would increase the number of lookup tables and associated interconnect needed to implement a function.

FPGA interconnect is comparatively expensive, both in terms of delay and area. An architecture that includes more long-distance connections would have faster interconnect, but the resulting chips might require more area for interconnect, reducing their logic capacity. Architectures with minimal interconnect resources will appear denser, but might be difficult to route. Architectures must address both integrated circuit and software goals.

The true capacity and speed of an FPGA is measured by the ability of design automation software to exploit the architecture. FPGA architectures and software must be developed simultaneously.

Software

The CAE industry has focused on the MPGA problem and has adopted a gate-like implementation model based on MPGA features. Many of the current software issues with FPGAs are a result of their non-gate-like implementation structure. This disagreement is most evident in the schematic entry library, which is a collection of gate-level primitives. The netlist generated from a schematic preserves the gate-like structure. The non-gate-like FPGA structure requires a partitioning step before the placement and routing process. Related problems in design automation have been addressed either as placement, considering only the physical constraints; or as technology mapping, considering only the logical constraints. Both sets of constraints must be solved simultaneously in order to produce implementations that are simultaneously dense and fast.

The partitioning problem is aggravated by the use of logic optimization algorithms originally designed for gate-like implementations. They often produce results that reduce speed and density rather than improve them. The reasons are varied, but traditional algorithms tend to factor logic aggressively, making more small gates; they ignore the ability of lookup tables to subsume larger amounts of logic. They also ignore routability considerations, which are of vital importance to FPGAs. New optimization algorithms are needed for lookup-table based FPGA architectures.

High-level synthesis and logic synthesis systems must target the high-level architectural features of FPGAs to gain the performance and density advantages they provide. The Library of Parameterized Macros (LPM) [Holley 1991] is an industry-sponsored standardization effort to develop an intermediate form that includes these high-level constructs. It may provide the appropriate interface between high level synthesis systems and systems-oriented FPGAs.

Placement and routing of FPGAs provides new challenges. The relatively slow FPGA interconnect structure demands true timing-driven placement and routing algorithms. Although these algorithms have been proposed for MPGA design automation, their usefulness for MPGA designs has not been great, and their adoption for FPGAs seems to be happening more quickly.

Partitioning in Space and Time

Because of the limited capacity of FPGAs, and their applicability to prototyping, FPGAs have re-kindled interest in multi-chip partitioning. There are several important problems that must be addressed, including FPGA resource estimation (logic, I/O and routing), timing and partitioning into dissimilar parts.

A farther-reaching problem is the issue of partitioning a design in time: identifying parts of a design that can be time-shared onto the FPGA, and generating separate FPGA configurations for them. At present, not only are there no algorithms, but the current design representations appear to be lacking in essential timing information. An elegant solution to this problem will allow true time-shared hardware and usher in a new era in hardware implementation.

Design Methodology

In an environment where the cost of prototyping is high, designers must rely on simulation to verify their designs. Highly-accurate simulators are very slow, while fast simulators gain speed at the expense of accuracy. An FPGA designer can replace extensive simulation with prototyping.

Reprogrammable FPGAs can be designed with a software-like iterative-implementation methodology (figure 2.5.1). The path from design to FPGA prototype is as short as a few minutes, allowing a designer to verify operation over a wide range of conditions more quickly and with more accuracy than simulation allows. Because the SRAM-based FPGA is reusable, there is no hardware cost for prototyping. The final prototype can become the production chip.

The tools and techniques required to support a prototyping-driven methodology are interesting areas of current work. Some of these tools exist already: the internal state of the SRAM-based FPGAs can be read back for verification. Design system features allow a designer to insert soft probes into the FPGA to examine internal nodes of a prototype during operation. The placement and routing algorithms can run in an *incremental* mode, adding logic with minimal impact on already-routed logic. Additional features may significantly improve the utility of the incremental design methodology.

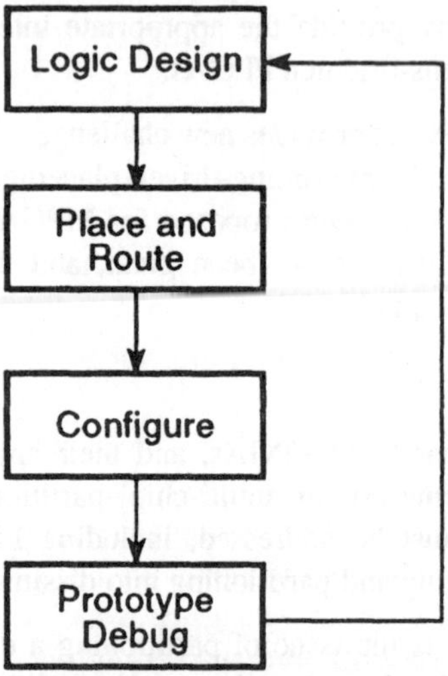

Figure 2.5.1. Incremental/Prototyping Methodology Design Flow.

2.6. Design Applications

This section begins with a discussion of general design principles that take advantage of the capabilities of SRAM FPGAs and shows how these principles are applied to canonical design examples such as state machines and counters. The discussion then covers design techniques that exploit the capabilities of FPGAs. The latter part of this section describes several designs and the techniques used to implement them. The designs demonstrate particularly effective use of SRAM FPGAs, especially with regard to the lookup tables and programmability.

General Design Issues

Design automation tools provide the designer with the ability to use an FPGA without knowing the details of the architecture. However, as with any implementation target, better understanding of the implementation and the operation of the software leads to more effective designs. This section describes some of the techniques that designers and design optimization software use to make effective use of SRAM-based FPGAs.

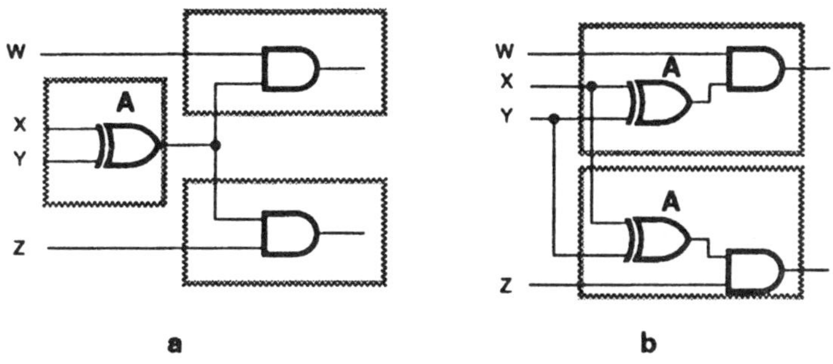

Figure 2.6.1. Logic and Partitioning (a) Before and (b) After Logic Duplication.

Duplicated Logic

Lookup tables have limited fan-in, but they implement any function of their inputs. A lookup table implements a four-input function with the same area and delay as a two-input function. A designer can take advantage of this capability by duplicating logic on the critical path. In figure 2.6.1, the shaded outlines indicate the logic that is mapped into a single lookup table. Since each lookup table has only one output, the A function must be in a separate lookup table from its destinations. In this example, duplicated logic reduces both the total number of lookup tables in the result and reduces the number of lookup tables along the path from inputs to outputs. The resulting logic is both denser and faster than the original.

One-Hot State Machines

In traditional discrete logic design, flip-flops are expensive. This is not true in the Xilinx FPGA architectures in which all combinational logic is followed by a flip-flop. The large number of flip-flops encourages one-hot encoding of state machines. In one-hot encoding, each state is represented by a separate flip-flop. A state machine with 16 states requires sixteen flip-flops, while a fully-encoded state machine requires only four. In PLD applications, where flip-flops are comparatively rare and there is no penalty for routing all state variables to all next-state calculations, this implementation would be woefully inefficient, consuming large numbers of macrocells for the flip-flops. However in an FPGA implementation, where lookup tables have limited fan-in, and where a flip-flop can follow any lookup table, the one-hot encoding often leads to a faster and smaller state machine.

One-hot encoding simplifies the next-state calculation because the states are already fully decoded. The decoded state can be gated with the transition equation for the next state in a single lookup table in one level of logic. In contrast, a complicated state encoding might require one level of logic to decode the state and a second to AND the state with the transition equation. The one-hot encoding saves one level of logic.

In addition, since the state bits connect only between states with transitions between them, and because state transitions tend to be localized, the routing of the state bits tends to be localized. The localized interconnect leads to shorter routes for the state bits, so interconnect delays are reduced.

Clock Enable

Gated clocks are known to cause problems in digital design because race conditions into the gate on the clock cannot be resolved before routing because precise delay estimates require knowledge of the routing paths of all signals. Xilinx provides clock enable signals on all flip-flops to remove the need for gated clocks. Rather than separate clocks, the design should include a single clock with the gating signal used as the clock enable on the flip-flop. This design eliminates potential glitches and allows the software to put the single clock on the dedicated global clock interconnect, which minimizes clock skew.

Performance Considerations

Since flip-flops on the FPGA are plentiful, pipelining has low cost. High-performance designs can be pipelined in single-function-generator stages that are followed by flip-flops. These techniques have resulted in designs in the XC3000 family that run at over 100 MHz, about 75% of the flip-flop toggle rate.

It is important to design at a high enough level to allow the FPGA to implement the logic effectively. The XC4000 architecture can implement a counter using the high-speed arithmetic support, but only if it is designed as a counter, not if it is designed as logic that implements the counter.

Iterative Design Methodology

Reprogrammability allows designers to experiment with the implementation technology with less effort and less expense than that required by board-level or MPGA prototypes. Designers can take advantage of reprogrammability by implementing the design one piece at a time. An effective design technique is to identify and implement performance-critical parts of a design first, lock that part in place and proceed to implement less-critical parts.

Counter Examples

This section describes different kinds of counters implemented in the XC3000/XC3100 to show a variety of density/performance points. These designs trade off counter features and density for speed. The addition or deletion of a single counter feature, such as parallel-load, can have a significant effect on the performance or density of the counter in the FPGA.

The counter designs in this section are shown at the gate level for descriptive purposes. In use, it is not necessary for an FPGA user to design counters at this level of detail. To incorporate a preferred kind of counter in a design, he or she may select the desired counter from a library.

The most straightforward way to build a counter is as a ripple-carry adder, one bit per XC3000 CLB (figure 2.6.2a). One lookup table implements the sum and the other implements carry. The counter value is stored in one of the flip-flops in the CLB. The carry chain passes through *n* CLBs in an *n*-bit counter.

A faster ripple-carry counter can be built with the same number of CLBs, but with the carry chain passing through *n*/2 CLBs. In this counter, the basic cell consists of two CLBs, shown in figure 2.6.2b. The first CLB implements two T-type flip-flops with independent data input for pre-load. The second CLB implements two bits of the ripple carry chain. Performance can be optimized by placement that allows the carry chain to use high-speed direct connect. A 16-bit counter using this design operates at 41 MHz in an XC3100-3 device

It is possible to build a non-loadable binary counter using less than one CLB per bit by segmenting the counter into three-bit pieces (figure 2.6.3). The least-significant tri-bit piece uses two CLBs to implement the first three bits of the counter as toggle flip-flops. One lookup table in the second CLB generates a parallel clock enable signal for the subsequent bits. The second tri-bit is also implemented in a pair of CLBs. Subsequent tri-bits require three CLBs since they use the parallel clock enable signals. A 16-bit counter using this design requires only 14 CLBs and can operate at 102 MHz.

A very fast non-loadable binary counter can be built by distributing the least significant bit with a shift register (figure 2.6.4). One CLB implements a one-bit "pre-scaler," halving the effective clock rate to the rest of the counter. Unlike traditional pre-scaling techniques, the clock signal is the same to all bits of the counter. A second CLB implements a two-bit counter that generates a parallel clock enable signal, CEP2, every eight clock cycles. Ripple carry can be used for the remainder of the counter. The Q_0 pre-calculation allows two clock cycles to distribute CEP2, which is adequate if it is distributed on a long line.

In order to distribute Q_0 quickly, it is replicated once for every bit in the counter. The copies are stored in CLB flip flops that are connected in a serial chain. When the counter starts, *n* cycles are required to initialize the chain of Q_0 copies. The duplication of the flip flops in the CLBs allows high-speed vertical direct connections to pass the value to the next CLB in the chain and horizontal direct connections to pass the counter bits. A 16-bit counter using this design takes 24 CLBs and operates at 204 MHz.

This section dealt with binary counters. For applications that can use non-binary counters, even faster designs are possible because they eliminate carry chain delays. These counters can be implemented in devices besides the XC3000, although dedicated carry logic in the XC4000 provides a superior method of implementing

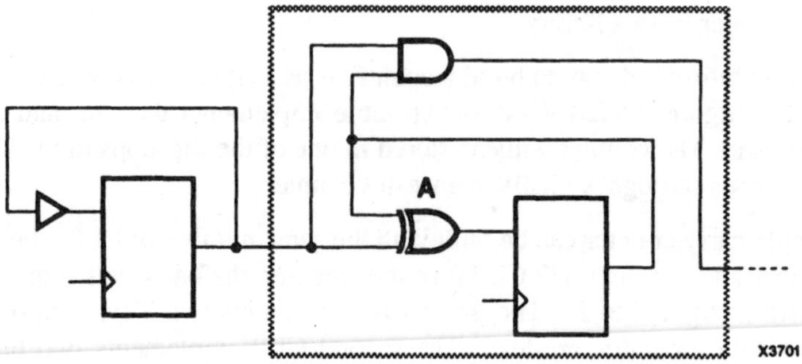

a

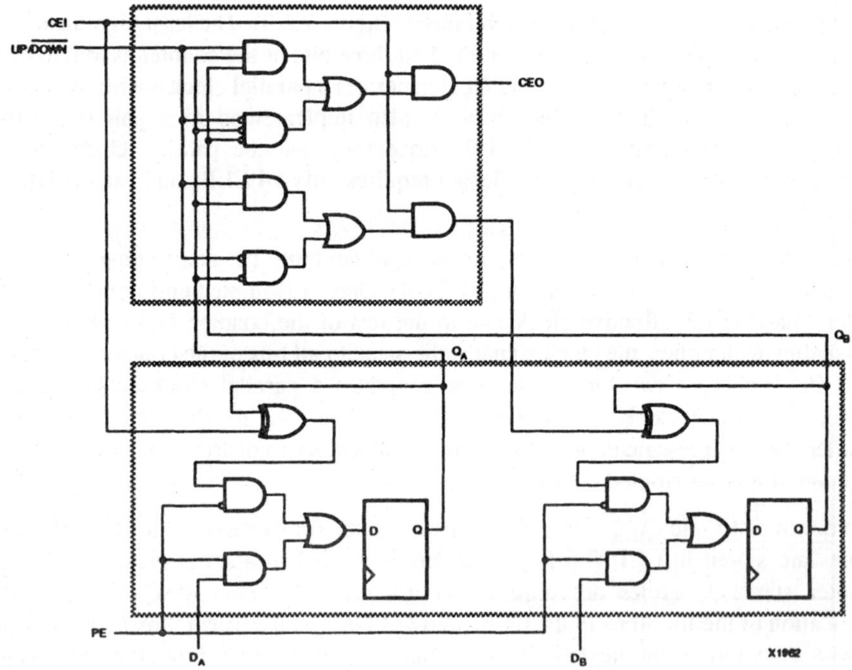

b

Figure 2.6.2. Ripple-Carry Counters. a) Simple Ripple-Carry Counter Built with One Bit per CLB. b) Faster Ripple-Carry Counter with One Bit per CLB.

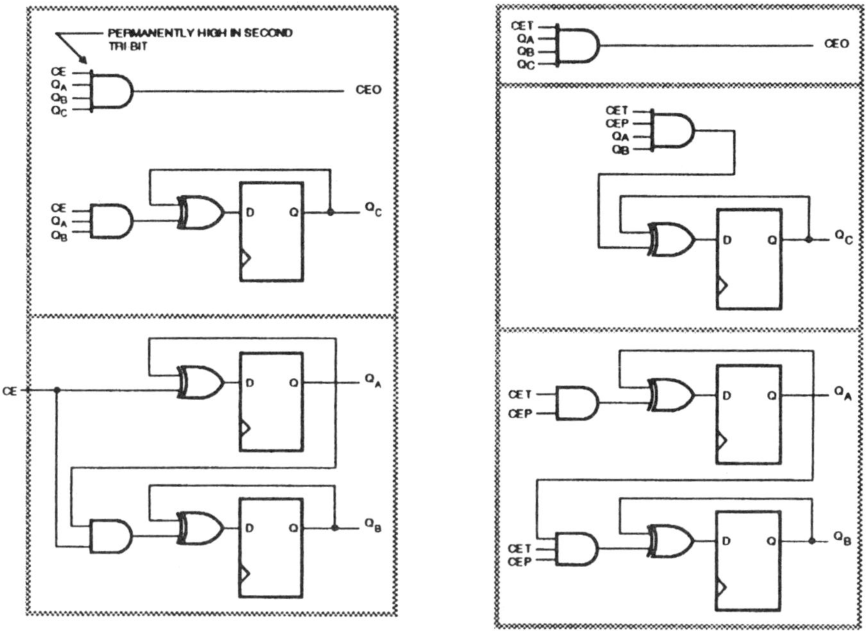

Figure 2.6.3. A Fast Binary Counter That Requires Less than One CLB Per Bit.

Table 4: Summary of 16-bit Counter Examples In an XC3100-3 Part

Type	Size (CLBs)	Speed (MHz)	Comments
Simple Ripple	16	23	loadable
Faster Ripple	17	41	loadable, up/down
Condensed	14	102	non-loadable
High-Speed	24	204	non-loadable initialization required

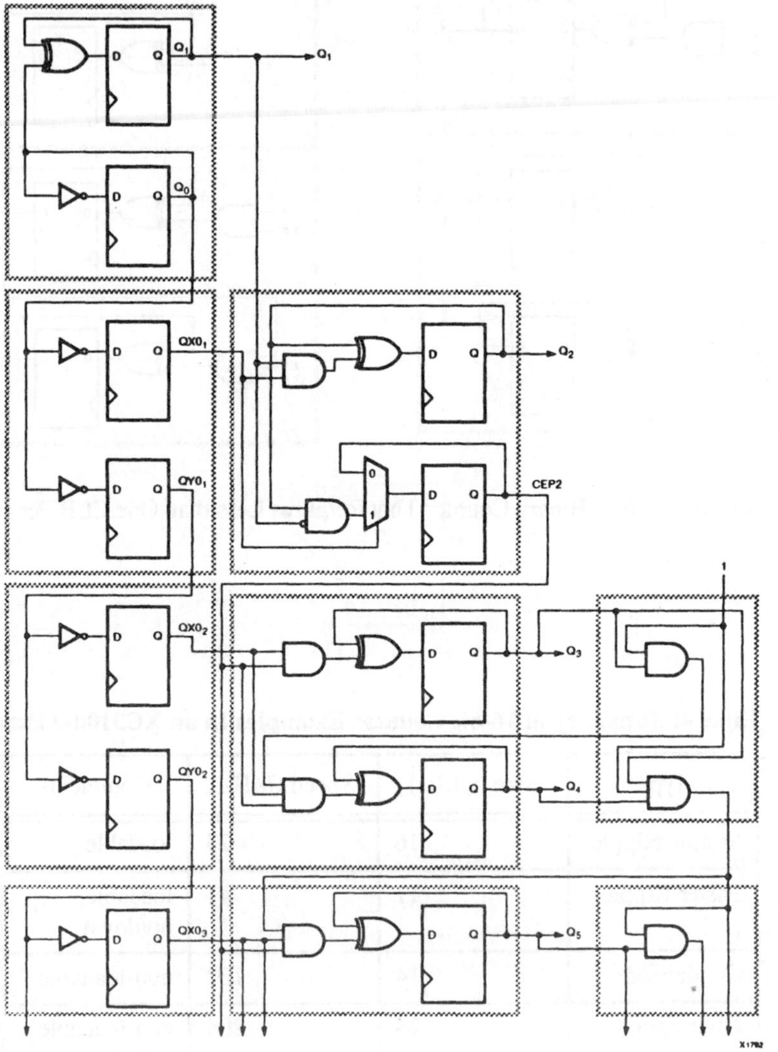

Figure 2.6.4. A Very Fast Synchronous Counter.

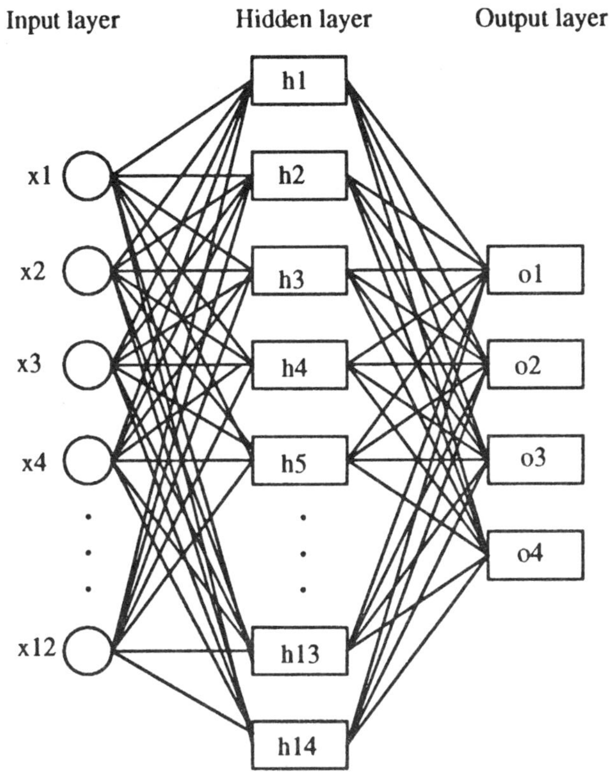

Figure 2.6.5. Ganglion Feed Forward Network (source Cox [1991]).

full-featured counters in that family. These techniques can be used to build faster counters with limited features. In the XC4000-5 speed parts, the high-speed counter runs at 111 MHz.

Efficient Multiplication by a Constant in an Artificial Neural Network

In an artificial neural network, the output of a neuron is calculated from the sum of its inputs multiplied by their respective weights. Cox [1991] used several XC3090 FPGAs to implement Ganglion, an artificial neural network for pattern matching. The Ganglion processor, shown in figure 2.6.5, implements a three-layer feed-forward artificial network. The first layer, the input layer, buffers the data from the twelve inputs. The second layer, the "hidden" layer, consists of fourteen units (neurons), each of which computes a weighted sum of the twelve inputs. The output layer computes a weighted sum of the fourteen units in the hidden layer.

Inputs are 8-bit unsigned values and the weights are 8-bit signed values. At each input to a unit, these two numbers are multiplied to produce a sixteen-bit product. The products are summed with a bias to produce a twenty-bit value which is scaled to eleven bits and passed to a large lookup table. The lookup-table implements the activation function, an arbitrary function of its input. The output of the lookup table is an eight-bit input to the next stage.

Each line connecting layers in figure 2.6.5 represents a multiplication of two eight-bit numbers. Clearly, an efficient implementation of this multiplication is crucial to the efficiency of this application. Since one of the numbers (the weight) is a constant, multiplication can be significantly simplified.

In the Ganglion design, the input eight-bit value is divided into two four-bit pieces, as shown in figure 2.6.6. Multiplication of a four-bit number by an eight bit constant produces a twelve-bit result. Each bit of the result can be expressed as a constant function of the four variable inputs bits. Therefore, each bit can be calculated in a single four-input lookup table in a single lookup-table delay. The twelve-bit result can be calculated with twelve four-input lookup tables. The two halves of the multiplication can be done in parallel, each in twelve lookup tables. The resulting partial products must then be added (with the high-order part offset by four bits). The eight overlap bits require eight full adders, the high-order four bits require four half adders to propagate the carry. The entire multiplication can be performed in thirty-two XC3000 CLBs.

In the Ganglion application, the multiplication constants may change from time to time as the network is re-trained to match new patterns. When this happens, the modification to the design is simple because only the lookup table contents change,

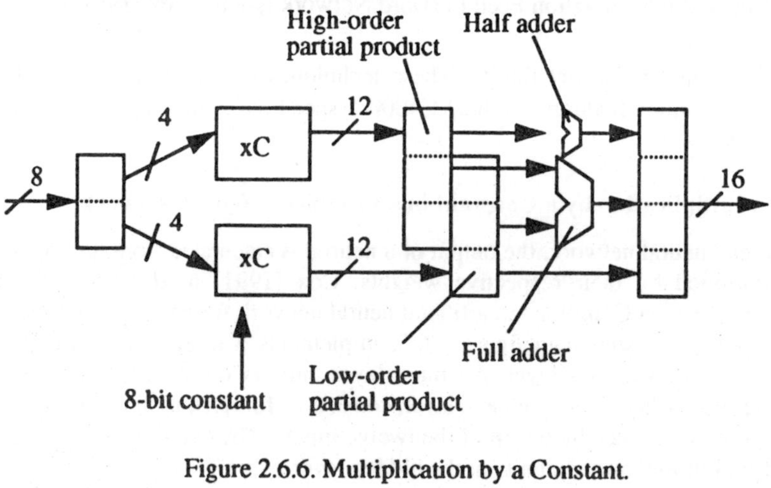

Figure 2.6.6. Multiplication by a Constant.

the interconnect does not change. The Ganglion design inserts new multiplication constants by modification of the lookup table contents in the configuration bitstream.

Distributed Arithmetic for Signal Processing

Distributed arithmetic is a method for implementing the sum-of-products that is the basis for many digital signal processing algorithms. Distributed arithmetic takes advantage of lookup tables and avoids a direct implementation of a multiplier, which would require a considerable amount of logic. This derivation is due to Mintzer [1992].

The basic sum-of-products expression for a digital filter is this:

$$Y = \sum_{k=1}^{K} A_k X_k$$

where K is the number of taps in the filter, the number of terms in the equation; and Y, A and X are multi-bit numerical values. The A_k are constants. Writing X as a sum of bits in fractional two's complement form, we get:

$$X_k = -x_{k0} + \sum_{i=1}^{B} x_{ki} 2^{-i}$$

where x_{ki} represents the ith bit of the kth number. If we substitute this expression for X into the filter function, expand the summations and re-group by powers of two, we get the following equations:

$$Y = \sum_{k=1}^{K} A_k \left(-x_{k0} + \sum_{i=1}^{B} x_{ki} 2^{-i} \right)$$

$$= -\sum_{k=1}^{K} A_k x_{k0} + \sum_{k=1}^{K} \sum_{i=1}^{B} x_{ki} A_k 2^{-i}$$

Expanding the summations and *distributing* the multiplication across the bits, we express the same function as follows:

$$Y = -(x_{10}A_1 + x_{20}A_2 + \ldots + x_{k0}A_k)$$

$$+ (x_{11}A_1 + x_{21}A_2 + \ldots + x_{k1}A_k)\,2^{-1}$$

$$+ (x_{12}A_1 + x_{22}A_2 + \ldots + x_{k2}A_k)\,2^{-2}$$

$$\vdots$$

$$+ (x_{1B}A_1 + x_{2B}A_2 + \ldots + x_{kB}A_k)\,2^{-B}$$

Written this way, the multiplications are all single-bit multiplications where the x_{ki} bits selectively include the A_k constants. This multiplication could be implemented as iterative selective addition, but Mintzer implemented it as a table lookup. Because the A values are constants, a table of multiple-bit entries can be pre-built so the summation of each row can be done in one lookup operation.

The lookup table has 2^K entries, one entry for each of the 2^K possible sums of the A_k. The value stored in the jth location in the table is the sum of the A_k for those bits that are 1 in the bit pattern of j. Each entry in the table consists of $B\log_2 K$ bits to preserve accuracy. The table is indexed by the x_{ki} for all k for one value of i. The table lookup implements the calculation in parentheses in the equation above. If K is small (for example, 4), the lookup table can be implemented in the lookup tables in the FPGA. The lookup table for large K requires external memory.

The final step in the distributed arithmetic calculation is the summation of the partial results, each offset by one bit position. This can be done serially, re-using the lookup table for all bit calculations, as shown in figure 2.6.7. The input variables, x_k are loaded in parallel into parallel-load shift registers (PLSR). The individual bits of the variables are shifted out and used as addresses for the lookup table in sequence (for subsequent values of i). As the bits are shifted, the same lookup table calculates the partial products for successive bit positions. The lookup table results are summed serially with one-bit shift on each cycle to offset the partial products. For signed arithmetic, the partial product from the sign bit is subtracted from the total.

In a DSP application, only the first register need be a parallel-load register, the others can be loaded serially from the output of previous stage.The serial paths are shown by the broken lines in figure 2.6.7.

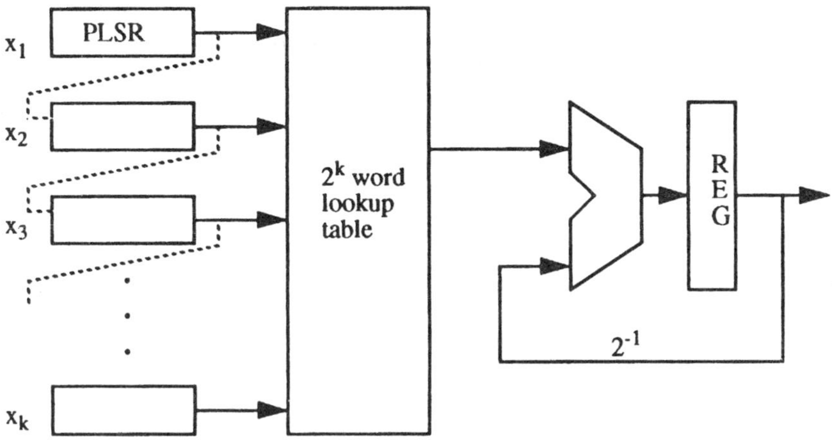

Figure 2.6.7. Distributed Arithmetic Sum-of-Products (source: Mintzer [1992]).

Mintzer built an 8-tap FIR filter using this technique. The 8-bit inputs are serialized and shifted as inputs to the lookup table. The lookup table contains only 16 6-bit words instead of 256 because of the symmetry of the filter response. The resulting filter uses 58 CLBs in an XC3042, and runs at a 1.4MHz data sample data rate.

Applications of Reprogramming

Xilinx FPGAs can be reprogrammed in 2ms to 30ms, depending on the part type. In most applications, the FPGA initiates its own programming automatically on power-up, and the application does not rely the reprogrammability of the parts. However, reprogrammability provides remarkably simple solutions to some common systems design problems.

Reprogramming for Board-Level Test

The most common system use of reprogramming is for board-level test and diagnostic circuitry. Since FPGAs are commonly used for logic integration, they naturally have connections to major subsystems and chips on the board. This puts the FPGA in an ideal location to provide system-level test access to major subsystems. The "operating" logic and the "test" logic need not operate simultaneously, so they can share the same FPGA, eliminating special-purpose test logic on the board. The system designer makes one configuration of the FPGA for normal operation and a separate configuration for test mode. The test configuration can be shipped with the board, so the test mode can also be invoked as a diagnostic after delivery without requiring external logic.

Rosendahl [1991] exploited reprogrammability for testing a bus interface built with a reprogrammable FPGA. One configuration of the FPGA is the "operating" logic, a bus interface; several additional configurations implement test circuitry. Test configurations of the FPGA include an IEEE 1149.1 boundary scan Test Access Port (TAP) for the board [Maunder 1990]. Specific tests are loaded with the TAP, including a *13N* pattern test generator for on-board RAM. The RAM tester generates addresses and patterns, captures and compares the results and keeps an error vector for later analysis.

Another test configuration of the same FPGA performs parametric tests. The test logic on the FPGA loads the external memory with a known value, then sends a read signal and address to the memory. It then samples the memory data lines after several different delays following the read strobe. The earliest match between the read data and the expected pattern gives the speed of the memory subsystem, which can be saved to become part of the system configuration. The resolution of the delay measurement can be the logic delay on chip, currently about 3ns, or it can be tuned with interconnect delays to be as fine as 1ns. Without a reprogrammable FPGA, the test and parameter measurement logic would have required thousands of additional gates of on-board logic as well as a significantly more complex board layout.

Configurable Interfaces

Test and diagnostic circuitry is a special case of a more general rule for the applicability of reprogramming: if a piece of logic can be split into multiple pieces that are not used simultaneously, then one FPGA can be time-shared to replace logic amounting to many times its maximum capacity. The system designer develops a separate FPGA program for each piece of logic, and builds the system to re-program the FPGA to switch between the pieces of logic as they are needed. This capability is especially useful in space-critical or weight-critical applications.

An opportunity to take advantage of reprogrammability occurs in interface design where a single system is required to interface with several different protocols. While any user of the system will use only one, the system manufacturer may wish to avoid the expense and complication of building a different piece of hardware for each protocol. The interface can be implemented in an FPGA which stores its programming in a PROM. A single PROM can contain multiple programs, and the choice of program selected during manufacturing or by the end user. This technique was used in Tellabs' channel interface card for the Crossnet 440 T1 multiplexer [Fawcett 1988]. The interface is implemented in an XC3030 FPGA, which can be configured to implement a Data Service Unit (DSU), Office Channel Unit (OCU) or secondary-mode OCU (see figure 2.6.8). Each T1 multiplexer system consists of four identical cards that can be configured independently. The end user selects the protocol

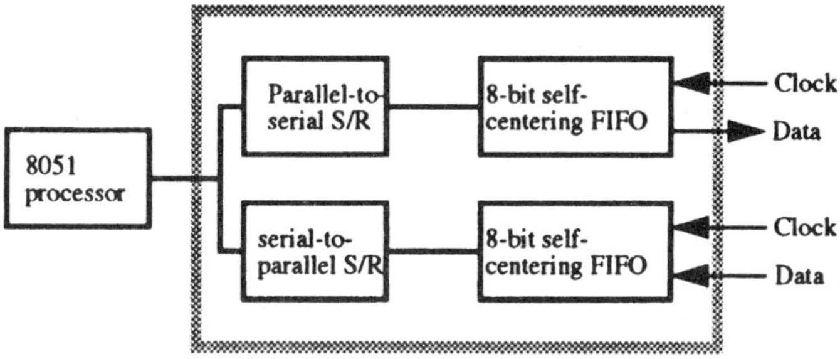

FPGA configured in DSU mode

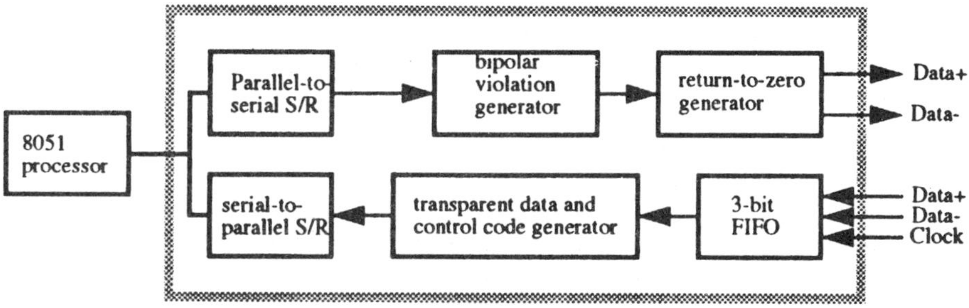

FPGA configured in OCU mode

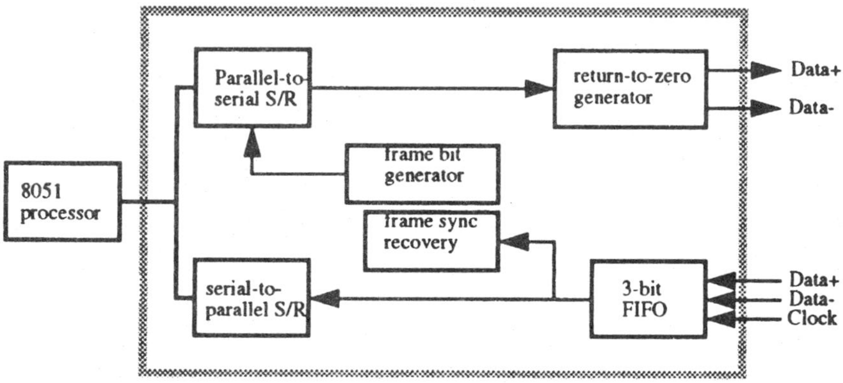

FPGA configured in Secondary OCU mode

Figure 2.6.8. Channel Interface Card with Different Interfaces Implemented in the FPGA.

for each card, and that selection determines the configuration that is loaded into the FPGA.

A similar technique was used by the Freeland Medical Division of Good Technologies in a video frame grabber for a PC-compatible computer. Seven FPGAs can be programmed to accept any one of the video formats used by different medical equipment. New video formats are easily added as new configurations for the FPGAs.

Hillen [1990] used an XC2018 FPGA to drive the printer interface of the Tektronix PhaserCard printer controller. The FPGA controls the flow of data from an external FIFO to the printer, formats the data, and handles signalling to the printer. Different configuration patterns for the FPGA allow the PhaserCard to control several different kinds of printers, including monochrome laser printers, ink-jet color printers, and wax-transfer color printers with parallel or serial interfaces.

When the board is idle and the printer is not printing, the FPGA is automatically reconfigured with logic to check printer status. An additional diagnostic configuration facilitates field testing.

Reconfiguration in Tape Drives

A tape drive cannot be reading and writing at the same time, so the logic for reading and writing can be implemented in the same FPGA. Liehe [1986] implemented the error correcting circuitry for a tape drive as two pieces: one for generating the error correction code when writing, one for checking the error correction codes when reading. The system reconfigures the FPGA when the tape drive switches modes. In another tape drive application [Fawcett 1993], seven FPGAs format data for different tape densities. Rather than supply different logic for each formatting style, the system programs the FPGAs to perform the proper formatting when the tape density is changed.

In this application, the FPGAs provided a second advantage: the system was originally shipped supporting only the most popular densities: 1600 and 6250 bpi. Later, the 800 and 3200 bpi formats were added and existing customers were updated in the field via floppy disk.

Reprogrammable Logic in a Configurable Display Device

The Radius Pivot monitor is a Macintosh-compatible display device that can display data either in portrait mode, preferred for word processing; or landscape mode, preferred for spreadsheets. A user of the display manually rotates the display to select the display style. Because the scan direction in the monitor is the same regardless of the orientation of the display, the display hardware must change the order of presentation of bits to the display to maintain Macintosh software compatibility.

The display interface board contains the frame buffers in video RAM, a Xilinx FPGA and the digital-to-analog converter for preparing the signal for display [Tan-Nguyen

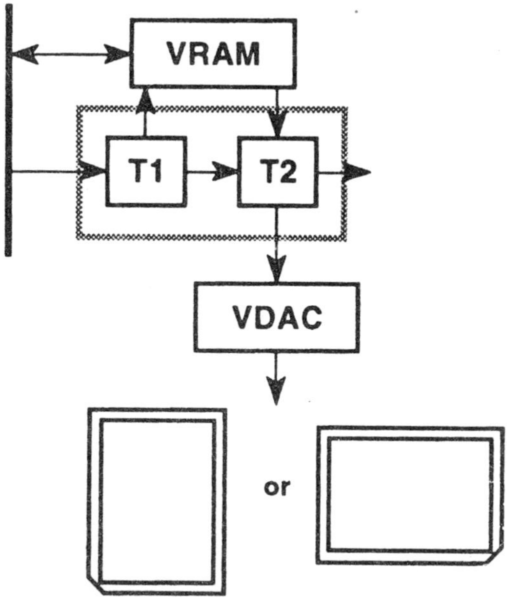

Figure 2.6.9. Pivot Display Structure.

1990] (figure 2.6.9). Part of the bitstream rotation is done when the pixels are stored in the display memory, part is done when the pixels are sent to the monitor. The data is scanned out of the video memory differently for portrait and landscape orientation. Therefore, the VRAM addresses generated by the FPGA are different in the different display modes. The final video stream goes to the monitor at 50 MHz.

The Pivot display actually has six formatting modes: one, two or four bits per pixel, with the display aligned either vertically or horizontally. These modes are all mutually exclusive, so one reprogrammable XC2018-100 FPGA with six different programs implements all six options. When the pivoting display orientation changes, an orientation-sensitive switch in the cabinet selects the correct program from a single PROM that contains all six programs and starts the reconfiguration process. Although XC2018 maximum capacity is 1800 gates, reprogrammability allows this part to replace about six thousand gates of logic.

A Fast Video Controller

Figure 2.6.10 shows a block diagram of a video controller for a full-page high-resolution display. The FPGA is required to control access to the video RAM, format the video data and generate control signals for the video monitor. Due to the high resolution, the system is required to run at a 70 MHz rate. The high throughput is

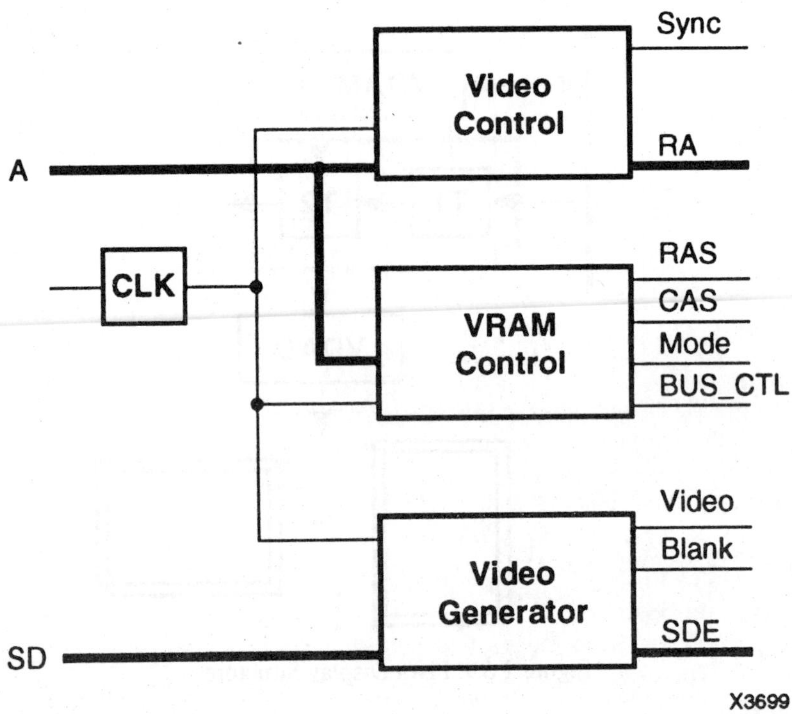

Figure 2.6.10. Seventy Megahertz Video Controller.

achieved by heavy pipelining, which is supported by the XC3000 CLB structure in which every logic block is followed by a flip-flop. To achieve the high performance, the logic was partitioned manually then placed and routed automatically. The placement and routing were done incrementally, with the few speed-critical paths placed and routed first and locked in place, before less speed critical paths were placed and routed.

This application was designed and debugged as successive prototypes in an XC3030 device, then put into production in an XC3020 device. The additional space in the XC3030 prototype device contained on-chip debugging logic, electronic "scaffolding."

A Position Tracker For a Robot Manipulator

Robot manipulator positions are tracked by counting the number of times an indicator has passed in front of a sensor. A manipulator with many degrees of freedom requires many sensors and position counters. The counters need not run fast, but they must be

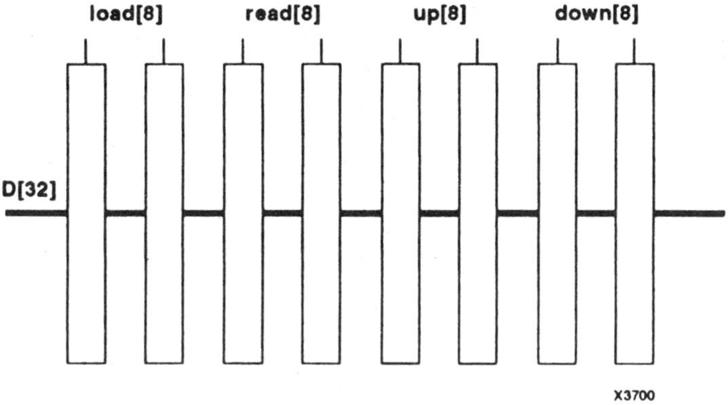

Figure 2.6.11. Robot Manipulator Position Tracker.

loadable; they must be able to count both up and down; and the system must be able to set and read the counts, the position of the manipulator, without interrupting the count.

The system in figure 2.6.11 controls a high-precision robot manipulator with 16 degrees of freedom. The range of motion of the arm and the position resolution require a 32-bit position register and counter for each degree of freedom. The design was implemented in two XC3090 FPGAs, each one implementing eight 32-bit loadable up/down counters. The position registers are in the XC3090's flip-flops, and the counters are implemented in lookup tables. Horizontal three-state lines give access to the position count registers. The position registers can be independently addressed to be read and written by the robot controller at a 30 MHz data rate. Each counter has its own up/down count signals and can count at 8 MHz. Each of the XC3090s contains over eight thousand gates of logic.

A Fast DMA Controller

Figure 2.6.12 shows a block diagram of a 16-channel DMA controller. The controller supports round-robin or priority channel selection with pipelined channel arbitration. It works with byte, word or long word transfers with separate 32-bit transfer count and address registers for each channel. The controller interfaces to 16-bit or 32-bit data busses and supports 20 million transfers per second, up to 80 MB/sec.

Figure 2.6.13 shows the block diagram of the DMA controller in an XC4008 FPGA. Registers and internal buffers use the CLB RAM configuration. Data and address sequencers use the high-speed arithmetic. Wide data busses run horizontally across the chip on three-state lines. The control logic for sequencing and DRAM interface is implemented as random logic on the same IC. The DMA controller represents approximately 7000 gates of logic and is implemented in about 200 CLBs,

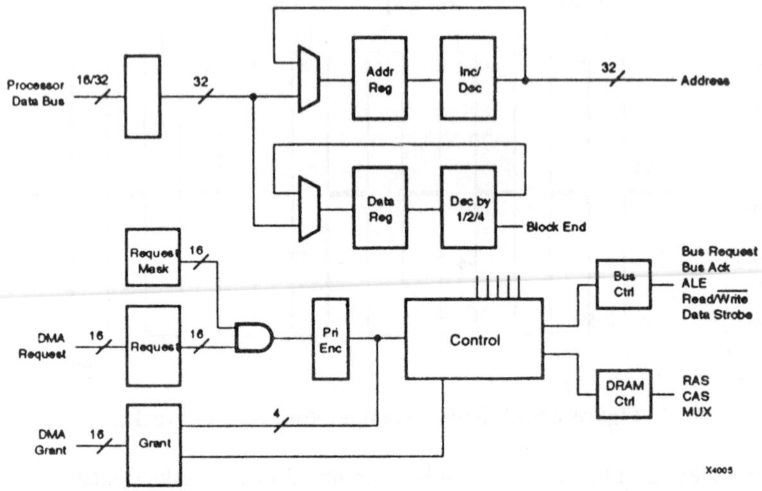

Figure 2.6.12. High-Speed DMA Controller Block Diagram.

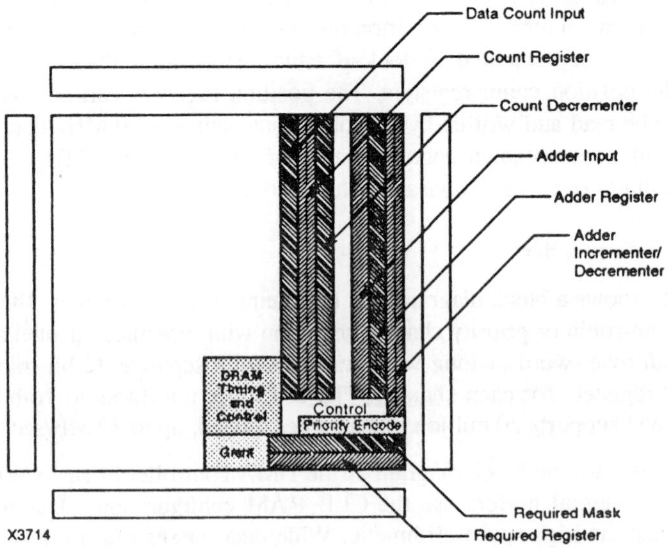

Figure 2.6.13. High-Speed DMA Controller Layout.

approximately two-thirds of the 18x18 array of CLBs in the XC4008 IC. The remaining area of the FPGA is available for other functions. A chained block scatter/gather option takes another 30 CLBs. If the DRAM is on the same board, the DRAM address multiplexer can be integrated on the XC4008 as well.

If the full functionality of this DMA controller is not needed, a simple 16-channel 32-bit DMA controller fits in only 72 XC4000 CLBs, and still runs at 20MHz.

Custom Computing Applications

SRAM-programmable FPGAs provide the opportunity to build a large-scale general-purpose system that can be reconfigured for new applications much like general-purpose microprocessors can be re-programmed with software to perform new functions. These systems have been classified as FPGAs for Custom Computing Machines, and are the subject of increasing interest as an alternative to supercomputers. This section gives an overview of a few of these systems and their innovative use of SRAM FPGAs.

The Quickturn Logic Emulator

Quickturn Design Systems, Inc. addressed the problem of prototyping large-scale ASICs and systems in the RPM Logic Emulator by using an array of Xilinx FPGAs [Walters 1990]. The RPM hardware includes a SCSI or Ethernet connection to a control processor, several emulation boards and external component adapters (figure

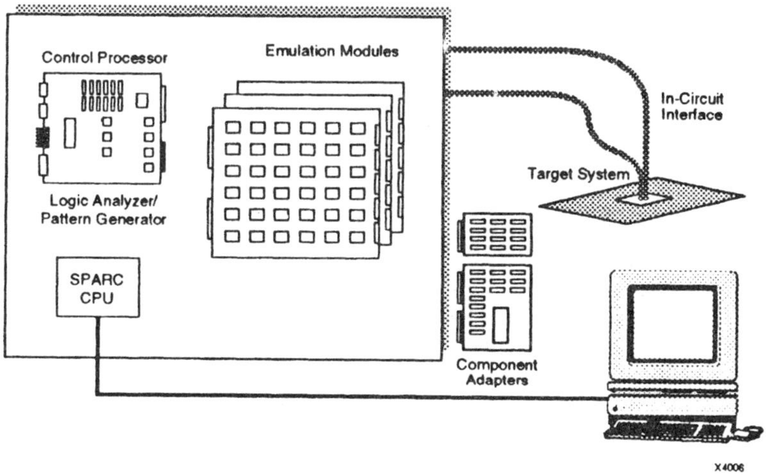

Figure 2.6.14. The RPM Emulation System.

2.6.14). Each emulation board includes an array of XC3090 FPGAs connected in a hypercube arrangement with fixed interconnects at the board level.

The external component adapters allow a designer to interface with external logic, either ICs or complete systems. The system includes configurable signal analyzers and generators, built with XC3090 FPGAs, that allow users to monitor the running system just as they would a simulation

The RPM software accepts a netlist of hundreds of thousands of gates, partitions the gates into FPGAs to optimize interconnection and density, places and routes designs on the FPGAs, and finally inserts buffers to adjust the signal timing to preserve the relative timing of the emulation with respect to the original netlist. The partitioner limits each FPGA to a small fraction of its logic capacity to ensure that enough interconnect remains in each FPGA to connect signals that pass through the design and to speed up the FPGA placement and routing that follows. Small changes are handled incrementally across all FPGAs in the system, limiting re-implementation times to minutes.

When partitioned and loaded on the emulator boards, an emulated design executes up to a million times faster than simulation.

Prototyping Hardware/Software Algorithms

Perle-0 [Shand 1989] is a platform for experimentation with logic and architectures for highly-parallel computation. The board consists of a 5x5 array of reconfigurable XC3020 FPGAs with local memory and VME bus interface circuitry to connect to a host processor. The FPGAs are connected in a two-dimensional array. Configuration for the entire array of chips is about four hundred thousand bits, with a download time of about 50ms. The array can implement functions up to about fifty thousand gates.

Perle-0 and its successor Perle-1, a 4x4 array of XC3090 FPGAs with 32 megabytes of RAM, have been programmed to perform a variety of algorithms, including image filtering, very-long-word-size arithmetic and RSA encryption. The time to implement the solutions to these problems is approximately the same as the time to implement a highly-optimized software solution to them. An algorithm is initially coded in a programming language for the host computer. The performance-critical part is re-coded into a model called a Programmable Active Memory (PAM), then the PAM model is mapped into the FPGAs. Changes in the algorithm may take weeks to describe and optimize, and design iterations can take several days. Compared to turnaround times for custom ASICs, these times are short, there are no tooling charges, and the hardware can be re-used.

Because the designers can continually improve their algorithms, Perle-0 and Perle-1 have achieved performance on several problems superior to the best custom IC solutions [Bertin 1992]. These include RSA encryption on 512-bit keys at 1500 bits per second and RSA decoding at 200 kbit/s, ten times faster than the fastest reported

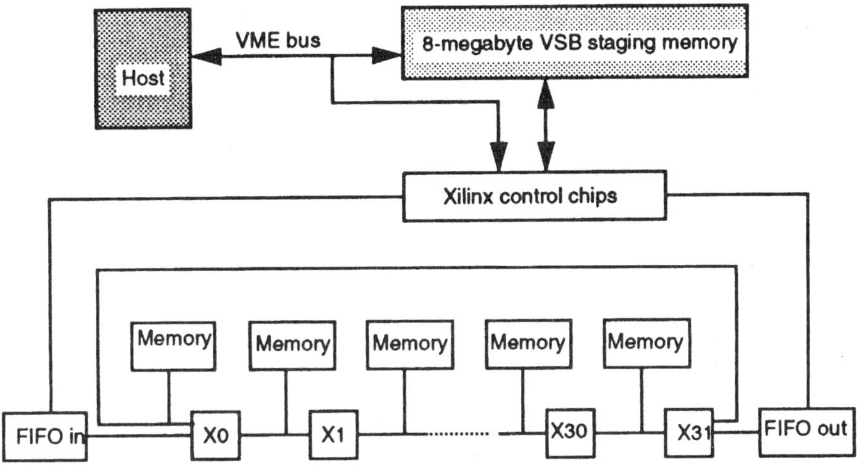

Figure 2.6.15. Splash 32-Stage Linear Array.

custom IC solution. A Perle-1 implementation of a finite difference method of computing solutions of the Heat and Laplace equations achieves performance comparable to a 25000 MIPS serial processor. A neural network emulator built on Perle-1 operates at 500 megasynapses per second.

Another example of this kind of application is the Splash processor [Gokhale 1991] [Lopresti 1991]. Splash consists of a linear array of function units (figure 2.6.15). Each function unit consists of a Xilinx XC3090 FPGA with a 128K byte external memory. The Splash processor is targeted to one-dimensional systolic problems and has been used for pattern matching DNA sequences. The general-purpose Splash board out-performs a Cray supercomputer on this problem by as much as a factor of 300 and out-performs a custom single-chip integrated circuit by a factor of 45.

Key to attaining high-quality designs in custom computing machines is the ability to repeatedly prototype the implementation, generating successively faster embedded designs, in much the same way that a software developer refines an algorithm by testing and monitoring. Since custom IC designers must eventually stop development and build the hard-wired design, they cannot continue to improve the algorithm.

A second advantage of reconfigurable systems is the ability to bring to bear large amounts of reusable hardware. Since the hardware is not dedicated to the application, is it cost-effective to bring dozens or even hundreds of FPGAs to bear on a single problem. This practice is prohibitively expensive when dealing with supercomputers, and custom ICs are typically not built to be scalable to multiple-chip solutions.

Custom computer designers build features into their systems to facilitate debugging and optimization of their highly-parallel designs. These debugging features access the readback facilities on the FPGAs to observe internal nodes. In some cases, readback is also used to read out the results of computations, saving the need to generate extra upload logic in the FPGA array.

There have been many different methods used to program custom computers. The programming methods are related to the intended applications. The RPM engine is programmed with a gate-level netlist that is partitioned among the FPGAs in the array. The Perle-board application developers use the XACT Design Editor and other editing tools for the PAM model. The Splash board uses VHDL [IEEE 1988] as its programming language. Other programming languages have been suggested, most notably a slightly-augmented C [Thomae 1991][Van den Bout 1992].

A Flexible Processor

Wolfe and Shen [Wolfe 1988] used several reprogrammable FPGAs to implement a "flexible processor." Eight reconfigurable FPGAs are used to implement instruction decoding, address generation and datapath operations in a single-board computer. They can be configured to implement a wide variety of virtual processor architectures with different instruction sets, addressing mechanisms, pipelining schemes and ALU operations. Wolfe and Shen used the flexible processor to prototype processor architectures to solve systems of linear equations.

Custom computing machines and flexible processors may change our definition of algorithmic complexity. For example, a searching algorithm that requires $O(n^2)$ operations on a standard processor may require only $O(n)$ time on a flexible processor configured with an array of comparators. The optimal choice of processor configuration may be dynamic and problem-size dependent. The processor configuration may even change during operation.

Reprogrammability gives a system designer new capabilities: reconfigurable hardware or even virtual hardware, similar to virtual memory in computer systems. Eventually such virtual hardware may be overlaid or time-shared in a manner analogous to multiple-process-management in computer systems, employing a software-like operational methodology as well as a software-like design methodology.

In a system with tight integration between the host and the custom processor, the host processor becomes redundant. If ever one is needed, the custom computer can emulate one.

2.7. Acknowledgments

I would like to thank all those whose work and innovations are described in this chapter. You are the ones who are pushing back the frontiers of knowledge. I would also like to thank Eddie Gutierrez and Jessica Fabula for their work with drawings; Lani Sutherland for research; and Bill Carter, Peter Alfke, Bernie New, Brad Fawcett,

Philip Freidin, Mon-Ren Chene, Jon Frankle, Erich Goetting, Danesh Tavana, Lee Salutos, Steve Kelem, Kerry Pierce, Laura Smith, Samiha Mourad and John Oldfield for reviews, suggestions and assistance with technical data.

2.8. References

P. Bertin, D. Roncin, J. Vuillemin, "Programmable Active Memories: Performance Measurements," *FPGA '92. First International ACM/SIGDA Workshop on Field Programmable Gate Arrays*, 1992.

M.A. Breuer, "Min-Cut Placement," *Journal of Design Automation and Fault Tolerant Computing*, October, 1977.

B.K. Britton, D.D. Hill, W. Oswald, N.-S. Woo, S. Singh, "Optimized Reconfigurable Cell Array Architecture for High-Performance Field Programmable Gate Arrays," *IEEE 1993 Custom Integrated Circuits Conference*, 1993.

W. Carter, K. Duong, R.H. Freeman, H.C. Hsieh, J.Y. Ja, J.E. Mahoney, L.T. Ngo, S.L. Sze, "A User Programmable Reconfigurable Gate Array," *IEEE 1986 Custom Integrated Circuits Conference*, 1986.

R. Cliff, B. Ahanin, L.T. Cope, F. Heile, R. Ho, J. Huang, C. Lytle, S. Mashruwala, B. Pedersen, R. Raman, S. Reddy, V. Singhal, C.K. Sung, K. Veenstra, A. Gupta, "A Dual Granularity and Globally Interconnected Architecture for a Programmable Logic Device," *IEEE 1993 Custom Integrated Circuits Conference*, 1993.

P. Chow, S.O. Seo, D. Au, T. Choy, B. Fallah, D. Lewis, C. Li, J. Rose, "A 1.2 μm CMOS FPGA using Cascaded Logic Blocks," *Proceedings of the Oxford 1991 International Workshop on Field Programmable Logic and Applications*, W.R. Moore and W. Luk, ed., Abingdon EE&CS Books, 1991.

C.E. Cox and W.E. Blanz, "Ganglion, A Fast Hardware Implementation of a Connectionist Classifier", *Proceedings of the Custom Integrated Circuits Conference*, 1991.

M.M. Denneau, "The Yorktown Simulation Engine," *Proceedings of the 19th Design Automation Conference*, 1982.

W.E. Donath, "Placement and Average Interconnection Lengths of Computer Logic", *IEEE Transactions on Circuits and Systems*, April 1979.

A.E. Dunlop, B.W. Kernighan, "A Procedure for Placement of Standard-Cell VLSI Circuits," *IEEE Transactions on CAD*, vol CAD-4, January, 1985.

C. Ebeling, G. Borriello, S.A. Hauck, D. Song and E.A. Walkup, "TRIPTYCH: a New FPGA Architecture," *Proceedings of the Oxford 1991 International Workshop on Field Programmable Logic and Applications*, W.R. Moore and W. Luk, ed., Abingdon EE&CS Books, 1991.

A. El Gamal, "Two-Dimensional Stochastic Model for Interconnections in Master Slice Integrated Circuits," *IEEE Transactions on Circuit sand Systems*, vol. CAS-28, no. 2, February, 1981, pp 127-138.

B. Fawcett, "Taking Advantage of Reconfigurable Logic," *High Performance Systems Programmable Logic Guide*, 1989. See also [Xilinx 1992].

C.M. Fiduccia, R.M. Mattheyses, "A Linear-Time Heuristic for Improving Network Partitions," *Proceedings of the 19th Design Automation Conference*, 1982.

R.J. Francis, J. Rose, K. Chung, "Chortle: A Technology Mapping Program for Lookup Table-Based Field-Programmable Gate Arrays," *Proceedings of the 27th Design Automation Conference*, 1990.

R.J. Francis, J. Rose, Z. Vranisec, "Chortle-crf: Fast Technology Mapping for Lookup Table-Based Field-Programmable Gate Arrays," *Proceedings of the 28th Design Automation Conference*, 1991. a.

R.J. Francis, J. Rose, Z. Vranisec, "Technology Mapping of Lookup Table-Based FPGAs for Performance," *IEEE International Conference on Computer-Aided Design*, 1991. b.

J. Frankle, "Iterative and Adaptive Slack Allocation for Performance-driven Layout and FPGA Routing", *Proceedings of the 29th Design Automation Conference*, 1992.

F. Furtek, G. Stone, I. Jones, "Labyrinth: A Homogeneous Computational Medium," *IEEE 1990 Custom Integrated Circuits Conference*, 1990.

M. Gokhale, W. Holmes, A. Kopser, S. Lucas, R. Minnich, D. Sweely, and D. Lopresti, "Building and Using a Highly Parallel Programmable Logic Array," *Computer*, January 1991.

S. Goto, "An Efficient Algorithm for the Two-Dimension Placement Problem in Electrical Circuit Layout," *IEEE Transactions on Circuits and Systems*, January, 1981.

J. Greene, V. Roychowdhury, S. Kaptanoglu, A. ElGamal, "Segmented Channel Routing," *Proceedings of the 27th Design Automation Conference*, 1990.

M. Hartoog, "Analysis of Placement Procedures for VLSI Standard Cell Layout", *Proceedings of the 23rd Design Automation Conference*, 1986, pp 314-319.

N. Hastie, R. Cliff, "The Implementation of Hardware Subroutines on Field Programmable Gate Arrays," *IEEE 1990 Custom Integrated Circuits Conference*, 1990.

S. Hauck, G. Borriello, S. Burns, C. Ebeling, "Montage: An FPGA for Synchronous and Asynchronous Circuits," *2nd International Workshop on Field-Programmable Logic and Applications*, 1992.

W.R. Heller, W.F. Mikhail, W.E. Donath, "Prediction of Wiring Space Requirements for LSI," *Journal of Design Automation and Fault Tolerant Computing*, May 1978.

D. Hill and N-S. Woo, "The Benefits of Flexibility in Look-up Table FPGAs," *Proceedings of the Oxford 1991 International Workshop on Field Programmable Logic and Applications*, W.R. Moore and W. Luk, ed., Abingdon EE&CS Books, 1991.

D.D. Hill, B.K. Britton, B. Oswald, N.-S. Woo, S. Singh, T. Poon, B. Krambeck, "A New Architecture for High-Performance FPGAs", *2nd International Workshop on Field-Programmable Logic and Applications*, IFIP, 1992.

K. Hillen, B. Fawcett, "Build Reconfigurable Peripheral Controllers", *Electronic Design*, March 6, 1990.

M. Holley, C. Kaplinsky, "Streamline programmable-logic design with the proposed LPM standard," *Electronic Design*, v 39, no 21, November 7, 1991, pp 89-96.

P. Horowitz and W. Hill, *The Art of Electronics, Second Edition*, Cambridge University Press, 1989.

H.C. Hsieh, K. Duong, J.Y. Ja, R. Kanazawa, L.T. Ngo, L.G. Tinkey, W.S. Carter, R.H. Freeman, "A Second Generation User-Programmable Gate Array", *IEEE 1987 Custom Integrated Circuits Conference*, 1987.

H.C. Hsieh, K. Duong, J.Y. Ja, R. Kanazawa, L.T. Ngo, L.G. Tinkey, W.S. Carter, R.H. Freeman, "A 9000-Gate User-Programmable Gate Array", *IEEE 1988 Custom Integrated Circuits Conference*, 1988.

H.C. Hsieh, W.S. Carter, J. Ja, E. Cheung, S. Schreifels, C. Erickson, P. Freidin, L. Tinkey, R. Kanazawa, "Third Generation Architecture Boosts Speed and Density of FPGAs", *IEEE 1990 Custom Integrated Circuits Conference*, 1990.

IEEE, *IEEE Standard VHDL Language Reference Manual, IEEE Std. 1076-1987*, 1988.

IEEE Computer Society Test Technology Technical Committee, *IEEE Std. 1149.1-1990. Standard Test Access Port and Boundary-Scan Architecture*, IEEE, New York, 1990.

K. Karplus, "Xmap: A Technology Mapper for Table-lookup Field-Programmable Gate Arrays," *Proceedings of the 28th Design Automation Conference*, 1991.

K. Kawana, H. Keida, M. Sakamoto, K. Shibata, I. Moriyama, "An Efficient Logic Block Interconnect Architecture for User-Reprogrammable Gate Array," *IEEE 1990 Custom Integrated Circuits Conference*, 1990.

T. Kean, *Configurable Logic: A Dynamically Programmable Cellular Architecture and its VLSI Implementation*, Ph.D. Dissertation, University of Edinburgh, Department of Computer Science, CST-62-89, 1989.

S. Kirkpatrick, C.D. Gelatt, Jr., M.P. Vecchi, "Optimization by Simulated Annealing," *Science*, 13 May 1983.

T. Liehe, "Two, Two, Two Chips in One," *Electronic Engineering Times*, November 17, 1986.

D. Lopresti, Rapid Implementation of a Genetic Sequence Comparator Using Field-Programmable Logic Arrays, *Advanced Research in VLSI: Proceedings of the 1991 University of California Santa Cruz Conference*, The MIT Press, 1991.

K. Lyons, "A Comparison of CMOS Static Random-Access-Memory Cells," *Electronic Engineering Times*, August 5, 1985.

C.M. Maunder, R.E. Tulloss, *The Test Access Port and Boundary-Scan Architecture*, IEEE Computer Science Press, 1990.

L. Mintzer, "FIR Filters with the Xilinx FPGA", *First International ACM/SIGDA Workshop on Field Programmable Gate Arrays*, 1992.

R. Murgai, Y. Nishizaki, N. Shenoy, R.K. Brayton, A Sangiovanni-Vincentelli, "Logic Synthesis for Programmable Gate Arrays," *Proceedings of the 27th Design Automation Conference, 1990.*

R. Murgai, N. Shenoy, R.K. Brayton, A Sangiovanni-Vincentelli, "Improved Logic Synthesis Algorithms for Table Look-Up Architectures," *IEEE International Conference on Computer-Aided Design*, 1991. a.

R. Murgai, N. Shenoy, R.K. Brayton, "Performance-Directed Synthesis for TAble Look Up Programmable Gate Arrays," *IEEE International Conference on Computer-Aided Design*, 1991. b.

H. Muroga, H. Murata, Y. Saeki, T. Hibi, Y. Ohashi, "A Large Scale FPGA with 10K Core Cells with CMOS 0.8 μm 3-Layered Metal Process," *IEEE 1991 Custom Integrated Circuits Conference*, 1991.

N.J. Nilsson, *Problem-Solving Methods in Artificial Intelligence*, McGraw-Hill, 1971.

P. Penfield, Jr., J. Rubenstein, "Signal Delay in RC Tree Networks," *Proceedings of the 18th Design Automation Conference*, 1981.

J. Rose, R.J. Francis, D. Lewis, P. Chow, "Architecture of Field Programmable Gate Arrays: The Effect of Logic Block Functionality on Area Efficiency," *IEEE Journal of Solid-State Circuits*, vol. 25, no. 5, October 1990.

J. Rose, S. Brown, "Flexibility of Interconnection Structures for Field-Programmable Gate Arrays," *IEEE Journal of Solid-State Circuits*, vol. 26, no. 3, March 1991, pp 277-282

G. Rosendahl, T. Paille, D. Freiling, R. McLeod, "In System Reprogrammable LCAs Provide a Versatile Interface for a DSP Based Parallel Machine," *Proceedings of the*

Oxford 1991 International Workshop on Field Programmable Logic and Applications, W.R. Moore and W. Luk, ed., Abingdon EE&CS Books, 1991.

P. Sawkar and D. Thomas, "Area and Delay Mapping for Table-Look-Up Based Field Programmable Gate Arrays," *Proceedings of the 29th Design Automation Conference*, 1992.

C. Sechen, A. Sangiovanni-Vincentelli, "The TimberWolf Placement and Routing Package," *IEEE Journal Solid State Circuits*, 20:510-522, 1985.

C. Sechen, "Chip-planning, Placement, and Global Routing for Macro/Custom Cell Integrated Circuits Using Simulated Annealing," *Proceedings of the 25th Design Automation Conference*, pages 73-80, 1988.

C. Sechen and D. Chen, "An Improved Objective Function for Mincut Circuit Partitioning," *IEEE International Conference on Computer-Aided Design*, 1988.

M. Shand, P. Bertin, J. Vuillemin, "Resource tradeoffs in fast long integer multiplication", *Proceedings of the 2nd Annual ACM Symposium on Parallel Algorithms and Architectures*, 1990.

J. Soukup, "Circuit Layout," *Proceedings of the IEEE*, October 1981.

J. Tan-Nguyen, T. Oyama, N. Moss, "Pivoting Monitor Increases Versatility of Workstations," *Computer Technology Review*, November 1990.

D. Thomae, T. Petersen and D. Van den Bout, "The Anyboard Rapid Prototyping Environment," *Advanced Research in VLSI: Proceedings of the 1991 University of California Santa Cruz Conference*, The MIT Press, 1991.

S. Trimberger, "Beyond Logic -- FPGAs for Digital Systems," *Proceedings of the Oxford 1991 International Workshop on Field Programmable Logic and Applications*, W.R. Moore and W. Luk, ed., Abingdon EE&CS Books, 1991.

S. Trimberger, "A Small, Complete Mapping Library for Lookup Table-Based FPGAs," *2nd International Workshop on Field-Programmable Logic and Applications*, 1992.

S. Trimberger, "A Reprogrammable Gate Array and Applications," *Proceedings of the IEEE*, July 1993.

S. Trimberger, M.-R. Chene, "Placement-Based Partitioning for Lookup-Table-Based FPGAs," *Proceedings of ICCD '92 International Conference on Computer Design, VLSI in Computers and Processors*, 1992.

D. Van den Bout, J. Morris, D. Thomae, S. Labrozzi, S. Wingo, and D. Hallman, "Anyboard: An FPGA-based, reconfigurable system," *IEEE Design and Test of Computers*, September 1992.

S. Walters, "Reprogrammable Hardware Emulation Automates System-Level ASIC Validation", *Wescon/90 Conference Record*, 1990.

A. Wolfe and J.P. Shen, "Flexible Processors: A promising application-specific processor design approach", Technical Report, Carnegie-Mellon University, 1988.

N.-S. Woo, "A Heuristic Method for FPGA Technology Mapping Based on Edge Visibility," *Proceedings of the 28th Design Automation Conference*, 1991.

Xilinx, *The Programmable Gate Array Data Book*, Xilinx, 1989, 1991, 1992, 1993.

Xilinx SMTO, "Static Memory Technology Overview,", Xilinx Technical Brief, Xilinx, 1991.

Chapter 3
Antifuse Programmed FPGAs

Dennis McCarty and Telle Whitney, Actel Corporation

3.1 Introduction

The architecture of an FPGA is determined, in large part, by the programmable switch technology used to configure it. Many such technologies have been considered for use in FPGAs, including laser programming [Smith] [Allen], pass transistors controlled by SRAM [Hsieh] [Carter] or EPROM cells [Wong] and antifuses [Gerzberg] [Hamdy] [Whitten].

This chapter describes the architecture, technology and use of FPGAs based on an electrically programmable two-terminal device known as an *antifuse* [Hamdy]. An antifuse device irreversibly changes from a high to a low resistance when a programming voltage is applied across its terminals. Antifuses offer several unique features for FPGAs, most notably their low *on* resistance of 100 to 600 ohms, and their small size. The layout area of an antifuse cell is generally smaller than the pitch of the metal lines it connects. It is about the same size as a via used to connect metal lines in a mask programmed array. More than 1,000,000 antifuses can now be integrated on a single FPGA, facilitating the development of routing architectures approaching the flexibility and scaling potential of conventional gate arrays. Current antifuse FPGAs offer complexity equivalent to an 10,000-gate conventional gate array and typical system clock speeds of up to 75 MHz.

While the focus here is on the Actel FPGA families similar principles will likely apply to other antifuse-based FPGAs now emerging. These developments include both new antifuse process technology and new FPGA architectures. The Act1[ElAyat] [ElGamal1] family ranges from 1200-2000 gates, the Act2 [Ahrens] family from 2500-8000 gates, and the Act3 [Schlageter] [Whitney] family ranges form 1500-10,000 gates.

Figure 3.1.1 shows a simplified block diagram of an ACT FPGA. Rows of logic modules are interspersed with horizontal routing channels containing predefined wiring segments of various lengths and offsets. Other wiring segments run vertically through the modules and across the channels. Each logic module computes a single-output function of several inputs. Each module input is connected to a dedicated

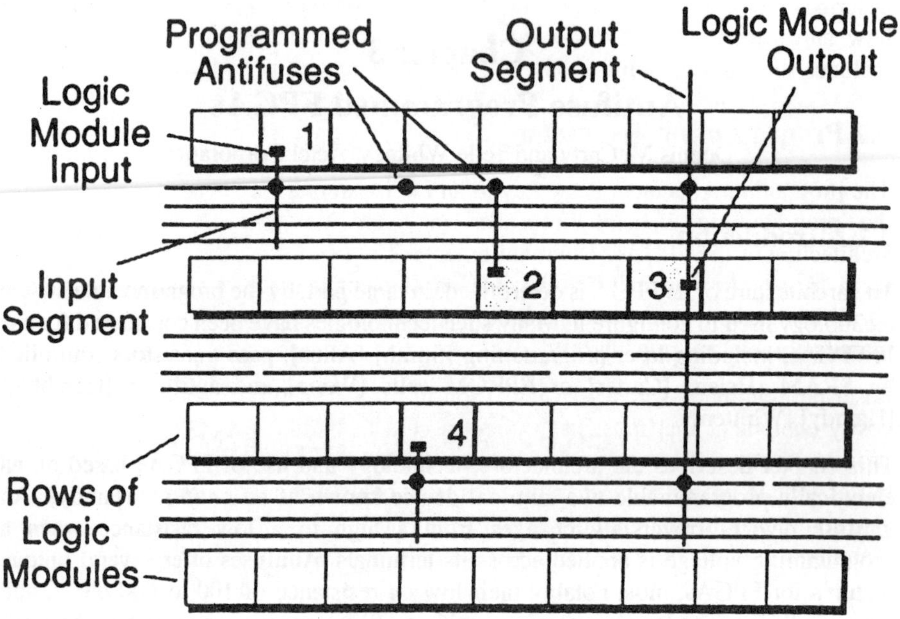

Figure 3.1.1. Actel Architecture

vertical wiring segment spanning either the channel just above or below the module. Each output signal appears on a dedicated vertical wiring segment of somewhat longer length. An antifuse is provided at each intersection of a horizontal and vertical segment, permitting them to be connected. The output of the driver, Module 3, in the figure, is connected to the Module 2 input by programmed antifuses to horizontal segments which, in turn, are connected to input segments. In the top channel, an antifuse is used to link two adjacent horizontal segments end-to-end, making it possible to reach an input of Module 1.

The central array of modules and channels is surrounded by input/output pads and buffers. Each IO buffer may be connected to the internal logic through a special module in an outer row or column of the array.

Succeeding sections of this chapter will consider the various factors that determine the speed, cost and ease of use of antifuse-based FPGAs. Section 3.2 describes the antifuse device, including its physical structure, manufactureability and reliability. Section 3.3 explains some basic principles of programmable routing applicable to any programmable switch technology, the *segmented routing channel* model Actel's routing architecture, and the basic architecture of Act1, Act2, and Act3. Section 3.4 describes the design flow and the design automation tools. Section 3.5 summarizes how antifuse FPGAs are expected to evolve in the future. Finally Section 3.6 includes some design applications.

3.2 Programming Technology

The programming element is the key to an FPGA architecture. The characteristics of the programming element influences in many fundamental ways the viable architecture of the FPGAs. The requirements of a programmable interconnect switch for a high-performance FPGA include an element that has a small area and low parasitic resistance and capacitance, as well as a switch technology that is manufacturable and reliable.

Laser-programmed switches [Allen] [Smith] offer decent performance, but require costly equipment that must have direct access to the un-packaged die in order to program the part. Although the switch itself is small, it often requires a surrounding buffer zone to protect adjacent structures from being damaged by the laser. There are also significant programming time and yield considerations which reduce production viability.

Early PROMs and PLDs employed electrically programmed fuses made of such materials as polysilicon, platinum silicide, tungsten-titanium, and nickel-chrome. These materials have proven to be both difficult to manufacture and program reliably for integrated circuits. The most common difficulty is that a programmed fuse can *grow back, or reconnect* over time if it was not programmed with an adequate current.

More recent architectures have used transistors as interconnect switches [Carter]. Although this approach is widely used, it has some significant costs. The appreciable resistance and capacitance of the switch transistor, and the large area of the SRAM or EPROM cell controlling it constrain the design of the routing architecture and the performance of the circuit.

Actel architectures employ a one-time programmable element the size of a via called an *antifuse*. An antifuse switch technology offers the following advantages for FPGAs:

- Antifuses have a significantly lower *on* resistance and parasitic capacitance than switch transistors, reducing RC delays in the routing.
- Antifuses are small, typically the size of via, and sit at the intersection of the horizontal and vertical routing wires.

Antifuses are off devices in an unprogrammed state. Their size allows the use of simple and flexible routing architectures with 1,000,000 antifuses on a 14100 (10,000 gate) device. Only a small fraction of the total number of antifuses need to be programmed, about 2% for a typical application. Antifuses fall into two categories: amorphous silicon and dielectric.

A layer of amorphous silicon placed between two metal layers undergoes a phase change when current is passed through it, becoming conductive. Devices based on this principle have been the subject of research for many years [Gerzberg] [Whitten] [Roesner] [Holmberg] [Lim] [Stopper], and were considered for an early FPGA design[Graham]. Their use has been hampered by two difficulties. First, application of a reverse current can return the amorphous silicon in a programmed antifuse to a nonconductive state. Second, even unprogrammed devices pass a small but significant current, termed *leakage current.* In a memory, where only a few bits must be active simultaneously, the problems can be avoided by careful design. The problems are more significant in an FPGA since the supply voltage is present across about half the antifuses at any given time.

Recent efforts at developing an amorphous silicon antifuse for FPGAs report the following results. Resistance is inversely proportional to the programming current, and is 50-110 ohms with a mode of 80 ohms at a programming current above 10mA [Birkner]. The capacitance contributed by each antifuse is 1.3 femtofarads in a 1.0 micron CMOS process [Birkner]. (This number does not account for the capacitance of the metal lines themselves, which contribute several times this amount per antifuse.) Pre-programming leakage current is under 10 nanoamperes at 5.5 volts [Whitten].

Dielectric antifuse consist of a single or multi layer of dielectric material placed between N+ diffusion and polysilicon. Upon application of sufficient voltage, the dielectrics breaks down. Early dielectric antifuses used a single-layer oxide dielectric. The remainder of this section focuses on the Programmable Low Impedance Circuit Element (PLICE), a multi-layer oxide-nitride-oxide (ONO) dielectric antifuse developed for use in FPGAs [Hamdy]. The PLICE is small enough that the area of a switch array is limited by the pitch of the metal wires rather than the size of the antifuse itself.

In the unprogrammed state the PLICE has a resistance over 100 giga-ohm. The PLICE can be programmed in one millisecond by applying 16 volts across its terminals. This programming pulse melts the dielectric, creating a conductive link of polycrystalline silicon between the electrodes. Typically, a single link is observed. The radius of the link increases with programming current, hence lowering the resistance. As the programming current flows, dopant atoms flow from both electrodes into the link, providing a controllable low resistance.

For a minimal area PLICE programmed with 5mA, the resistance is distributed around 600 ohms as shown in Figure 3.2.1 The resistance distribution is significantly

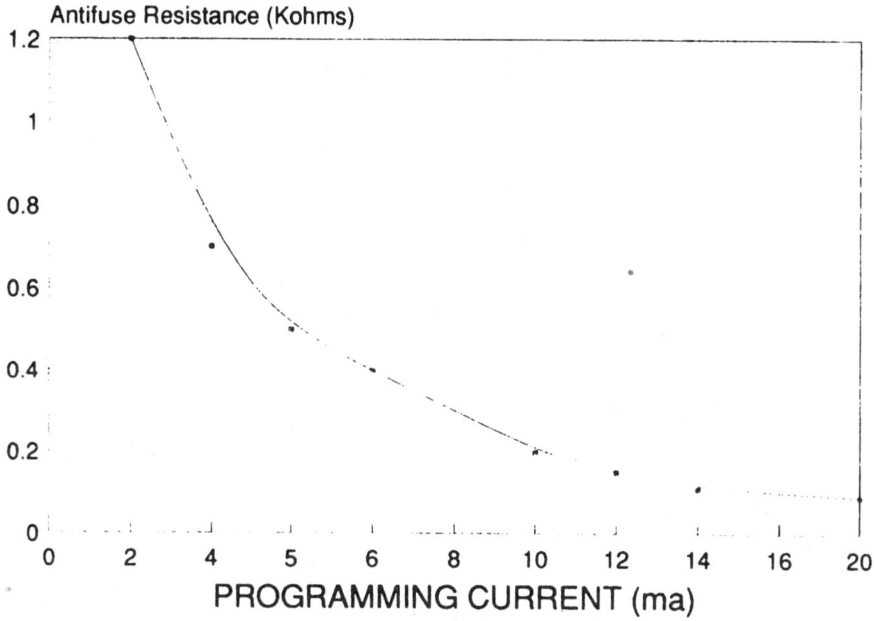

Figure 3.2.1. Resistance of Programmed Antifuse versus Programming Current

tighter than obtained with simple oxide dielectrics. With a larger cell layout that reduces the parasitic resistance to the electrodes, and higher programming currents of 14 mA, the distribution moves down to about 100 ohms, providing an area-delay trade-off. Use of ONO also improves both the yield and the reliability compared to oxide antifuses [Chiang1]. Capacitance is 6 femtofarads per antifuse in a 0.8 micron CMOS process; this includes the contribution of the polysilicon and diffusion electrodes and the metal lines used to connect them [Chen]. An unprogrammed PLICE has a leakage current of about one femtoampere, so even for the largest FPGAs the total leakage is negligible.

Use of the PLICE adds three masks to a conventional double-metal CMOS process. It can be fabricated in a typical CMOS facility using standard material, processing equipment and techniques.

A PLICE consists of an oxide-nitride-oxide structure sandwiched between N+ diffusion and N+ polysilicon gate, as shown in Figure 3.2.2 . A thin layer of oxide is thermally grown on top of the N+ surface, followed by LPCVD nitride and the reoxidized top oxide. Finally, the top polysilicon electrode is implanted with arsenic.

Antifuse reliability must be considered for both the unprogrammed and programmed

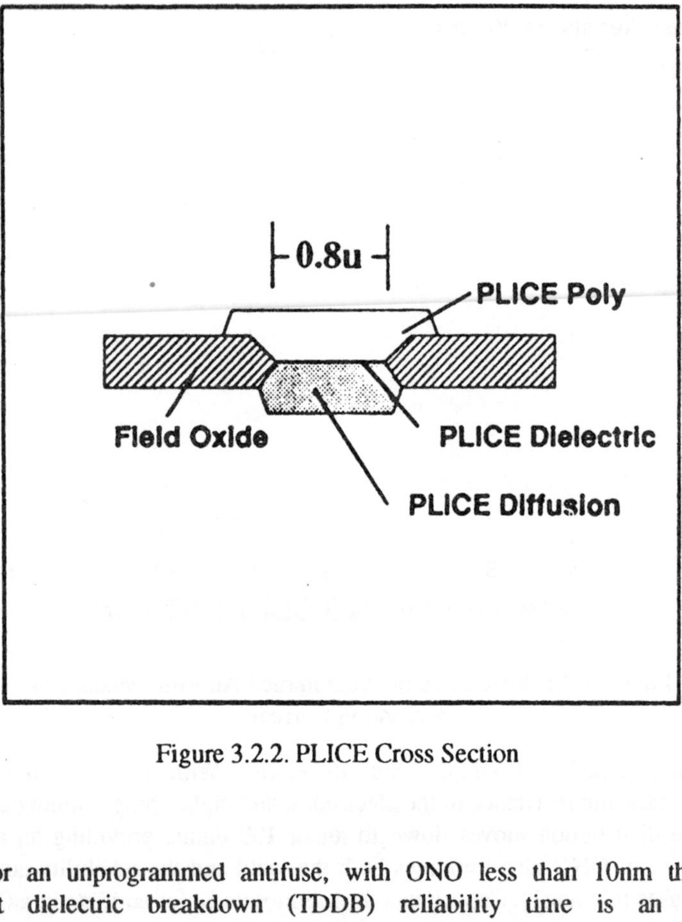

Figure 3.2.2. PLICE Cross Section

states. For an unprogrammed antifuse, with ONO less than 10nm thick, time dependent dielectric breakdown (TDDB) reliability time is an important consideration. Ordinary accelerated testing using electrical field and temperature stress was done in order to extrapolate the dielectric's lifetime under normal operating conditions. Figure 3.2.3 shows a plot of the time-to-breakdown vs. the reciprocal of the electric field applied to the dielectric, which has been shown to be an appropriate model [Chiang1]. Based on this data, one may extrapolate a lifetime for an ONO antifuse of well over 40 years of normal operation at 5.5V and 125C.

It is equally important that the resistance of a programmed antifuse remain low during the life of the part. Single-layer oxide dielectrics are known to be susceptible to *self healing* where resistance increases over time. Such increases do not occur for ONO dielectrics. Temperature-accelerated measurements reveal no intrinsic failure mechanism; the programmed antifuse resistance remains unchanged in all cases. The true lifetime of a programmed antifuse has yet to be determined since normal CMOS electromigration failures destroy the test structure first.

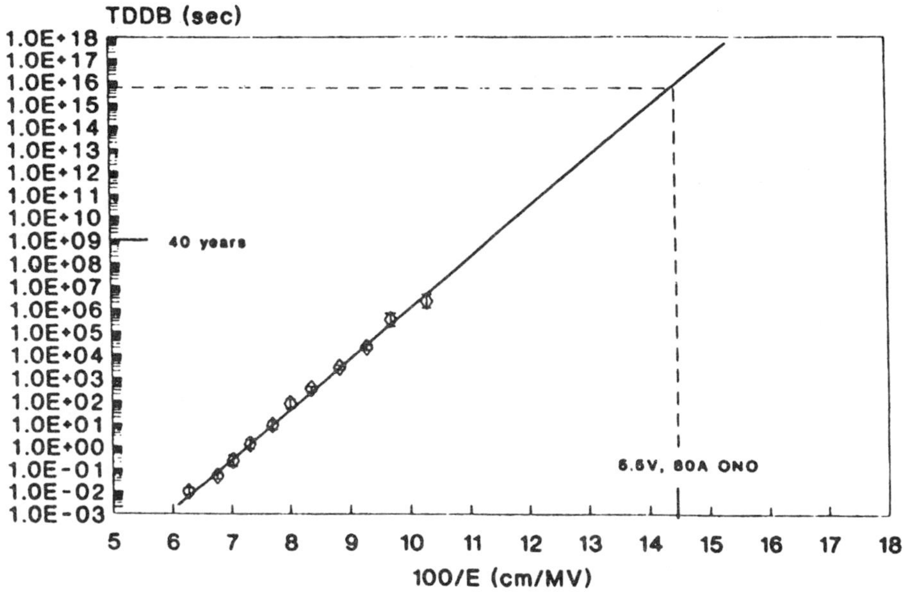

Figure 3.2.3. ONO Reliability, 1/E Model (95% Confidence Intervals)

Of course, the reliability of the FPGA is affected by the reliability of the base CMOS process as well as the PLICE. PLICE-based FPGA product reliability studies show the same failure levels encountered with normal CMOS circuits [Chiang2]. The ONO dielectric is highly radiation resistant. Initial results show that products containing ONO antifuses can withstand 1.5 million rads.

3.3 Device Architecture

This section describes in detail the Actel antifuse FPGA architecture. The first section is an overview of the architecture. Subsequent sections describe Act1, Act2 and Act3 focusing on their differences.

Principles of Programmable Routing

A routing architecture for an FPGA must meet two criteria: routability and speed. Routability refers to the adaptability of the wiring segments to accommodate all the nets of a wide variety of applications. Only the switches connecting the wiring segments may be customized (by programming) for a specific application, not the numbers, lengths or locations of the wiring segments themselves. While sufficient wiring segments for good routability must be provided, excess wiring segments waste chip area.One of the important architectural considerations is that the routing of an application can be determined by an automatic routing program with little or no manual intervention required.

Propagation delay through the routing is the other major factor in FPGA performance. In any gate array architecture, whether mask or field programmable, it is inevitable that some nets require longer routings than others. After routing, the exact parasitic resistance and capacitance of the wiring are known and the delay on each net can be computed accordingly. The resulting *post-layout* net delays will vary according to some statistical distribution. If the average of this distribution is high, it will limit the performance of the design. If the distribution is too broad, a user will have difficulty estimating delays for the application.

The delay distribution for mask-programmed arrays is narrow from the low interconnection impedance so that variation isn't a major difficulty for the designer. The problem is more challenging for FPGA architectures. Any programmable switch (EPROM, MOS pass device, or antifuse) has a significant resistance and capacitance product (RC). Each time a signal passes through a programmable switch, another RC stage is added to the propagation path. For a fixed R and C, the propagation delay mounts quadratically with the number of RC stages in series. This increases the average of the net delays and broadens the post-layout delay distribution. The use of a low resistance switch, such as the antifuse, limits RC to keep the average delay low and its distribution tight.

Of equal significance to the performance advantages of the antifuse is optimization of the routing architecture that it affords. Some of the trade-offs between the length of wiring segments in a channel, the area required by the segments, and the resistance and capacitance of the switch are illustrated in Figure 3.3.1 .

Figure 3.3.1 (a) illustrates a set of nets routed in a conventional masked device channel. With the complete freedom to configure the wiring afforded by mask programming, the positions and lengths of the horizontal wires may be customized for the particular set of nets. The *left edge algorithm* [Hashimoto] shows how to do the customizing using a number of tracks equal to the *channel density*. The channel density is defined as the maximum number of nets passing through any cut across the channel [Lorenzetti]. This figure assumes there are no *vertical constraints* [Lorenzetti], since they do not occur in FPGAs where each signal enters or leaves the channel on its own vertical segment.

In an FPGA, achieving the freedom of an unconstrained channel would require switches at every cross point, as shown in Figure 3.3.1 (b). These switches between adjacent cross points along a track allows the track to be subdivided into segments of arbitrary length. Since the number of RC stages encountered by a net is proportional

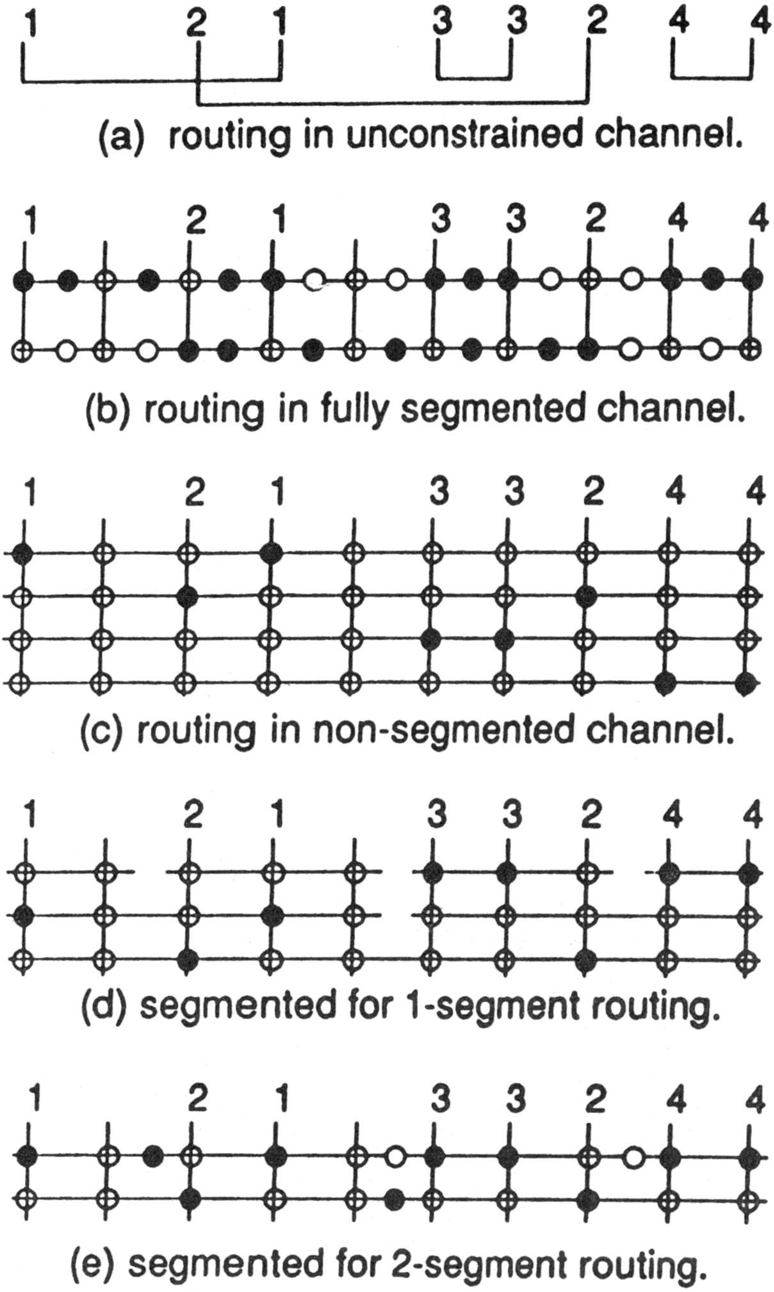

Figure 3.3.1. Illustration of Antifuse Routing

to its length, the delay of long nets becomes unacceptable.

Another alternative would be to provide a number of full length continuous tracks large enough to accommodate all nets, as shown in Figure 3.3.1 (c). This approach is used in the switch arrays of many types of programmable logic arrays and certain programmable logic devices (e.g. [Wong] [Marr]). The advantages of this model are that only two RC stages are encountered on any net, and that the delays of the nets are identical and predictable. However, even short nets incur the capacitance of a full track length. Furthermore, the area required is excessive and grows quadratically with the number of nets.

A segmented routing channel offers an intermediate approach. The tracks are divided into segments of varying lengths (Figure 3.3.1 (d)), allowing each net to be routed using a single segment of the appropriate size. Greater routing flexibility is obtained by allowing multiple adjacent segments in the same track to be joined end-to-end by switches (Figure 3.3.1 (e)). Enforcement by the software of simple limits on the number of segments joined or their total length guarantees that the delay will not be unduly increased by joining segments.

The problem of routing a segmented channel, or assigning a segment s to each unique net, is solvable in linear time for single-segment routing, Single-segment routing is illustrated by the model of Figure 3.3.1 (d). When a net is allowed to use multiple segments, as illustrated in Figure 3.3.1 (e), the general routing problem becomes more difficult (in particular it is NP-complete [ElGamal2]). However, many important special cases can be solved in polynomial time, and practical applications can be routed in a few minutes on a personal computer using heuristic methods. Reference [ElGamal2] gives a more thorough review of algorithms for segmented channel routing.

How does one design a segmented channel? In a conventional masked device channel the number of tracks, or *channel width*, must be chosen to accommodate most applications. Statistical models have been developed to estimate the required channel width (e.g. [ElGamal3]). In a segmented channel device, both the channel width and the segment lengths and offsets must be chosen carefully to suit the statistics of anticipated applications. Remember these applications are varied, and may have quite different routing requirements. Analytical solutions to the problem are as yet unknown, but experience indicates that even channels designed in an ad-hoc manner can be quite efficient (with limited use of multiple-segment routing).

Surprisingly, a well-designed segmented channel does not require many more tracks than would be needed in a conventional channel. It is an interesting finding, given the considerable restrictions segmented routing imposes, but it is supported both experimentally and analytically. A distribution giving the probability of occurrence of a connection as a function of length and starting point was derived from placements of 510 channels from 34 applications. Two segmentations were designed for a channel with 32 tracks: one intended for routing using exactly one horizontal segment per net, the other for routing using one or two horizontal segments per net. Sets of nets with various densities were chosen randomly from the distribution, and an attempt was made to route each set in both channels.

The results of this test may be seen in Figure 3.3.2 . In a conventional channel any

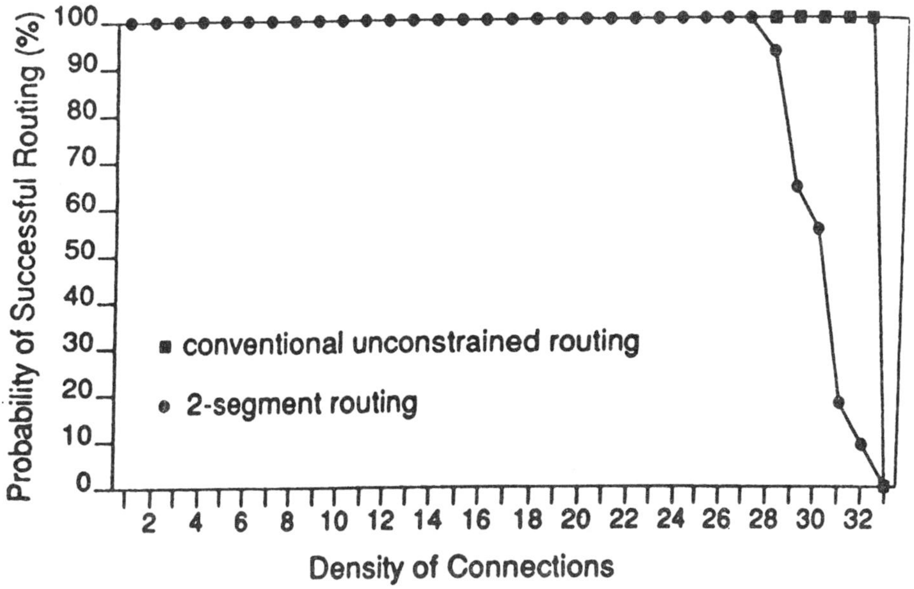

Figure 3.3.2. Results of Routing of 520 channels from 34 applications

design containing routing channels with density equal to 32 or less can always be routed. Allowing two-segment routing provides results only a little worse than the unconstrained case. A high probability of routing is observed when the density is only three or four below the number of tracks. (Further details of this study are given in [Greene1]).

An asymptotic analysis [ElGamal2] has confirmed that the number of tracks required in a segmented channel grows linearly with the expected channel density, just as with conventional channels. This is true even for single-segment routing.

Routing Architecture of the Actel FPGAs

We now describe in detail how segmented routing is applied in the ACT FPGA routing architecture. This section presents a general view of the Actel architecture and is followed by sections describing Act1, Act2, and Act3 Figure 3.3.3 shows a

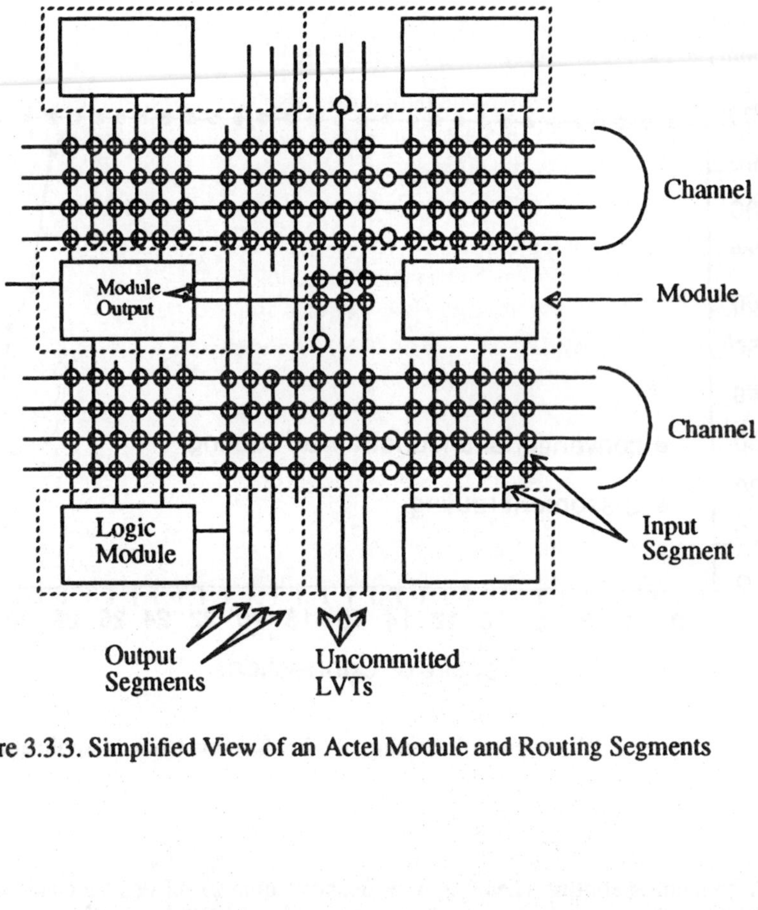

Figure 3.3.3. Simplified View of an Actel Module and Routing Segments

simplified schematic view of a module and the adjacent wiring segments and antifuses. Two segmented routing channels extend horizontally above and below the module. The input segments each span only one channel in order to minimize their capacitance and the total wiring area. Each input segment can be connected to any of the uncommitted horizontal segments in its channel through an antifuse. Other antifuses connect horizontal segments end- to-end to support multiple-segment routing. Each channel also contains one full length segment that is grounded and another tied high so that any input can be programmed to logical 0 or 1. The figure shows six inputs per module, but the actual number varies according to the module function; it is typically eight for a basic logic module.

Output segments span two channels above and two channels below the module. Segments reaching the top or bottom channel may be slightly longer or shorter. Additional uncommitted vertical segments of varying lengths are also provided. They may also be joined end-to-end to form vertical segmented channels. (In the actual layout of the device all vertical segments pass over the modules and the term *channel* refers only to the region occupied by the horizontal segments.)

In Act2 and Act3 devices, a module output may also access the uncommitted vertical segments directly through one antifuse, called an *F antifuse*, outside the channel. Each module has access to the uncommitted segments on either of its sides allowing two modules to share the same set of *Long Vertical Tracks (LVT)*.

The routing shown in Figure 3.1.1 is typical of most nets. In order to minimize the number of series RC stages, each channel's horizontal segments are driven directly by the dedicated output segment. This style of global routing is classified as a *Steiner Tree with Trunk*. [Preas].

Although the favorable routing of Figure 3.1.1 , can be assured for speed critical nets, some 5-10% of the other nets must be placed with a module input in some channel beyond the reach of the dedicated output segment. In this case, a suitable uncommitted vertical segment is selected to provide an alternate trunk for the channels not reached by the output. The vertical segment is driven directly from the output through an F antifuse, as shown in Figure 3.3.4 . Since this antifuse may be called upon to drive nearly the whole capacitive load of a widely dispersed net, the delay is very sensitive to its resistance. By programming these antifuses with higher than normal currents and providing them with additional strapping contacts their resistance is reduced to about 100 ohms, compared to the 500 ohms of a typical antifuse. The *F* antifuses thus require greater layout area but since there are few of them the cost is negligible.

In the rare instance where an uncommitted vertical segment is needed, but none is directly accessible to the module output, a segment from some other column must be used. In this case the uncommitted segment is driven through one of the horizontal channels spanning itself as well as the module output as shown in Figure 3.3.5 . The route involves an extra RC stage, so it is reserved for nets whose speed is not critical. Assignment of uncommitted segments in each horizontal and vertical channel to those nets that need them constitutes the segmented channel routing problem. Algorithms for solving these are discussed in [ElGamal2].

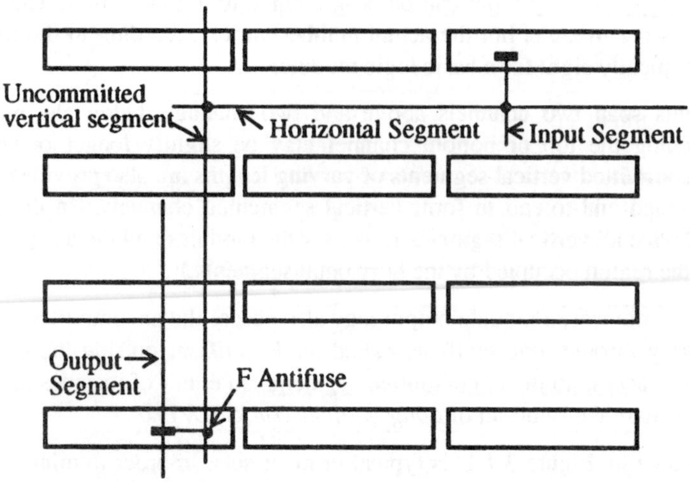

Figure 3.3.4. Routing Using Long Vertical Track (LVT)

Act1 Architecture

FPGA architectures can be characterized by the complexity or granularity of their basic logic module. A simpler module has lower internal delay and, since modules require less area, more of them can be provided on the chip. Furthermore, a *fine-grained* architecture tends to be more flexible. For example, a wide variety of functions may be built with equal efficiency out of two-input NAND gates, but eight-input NAND gates are much better at some functions than others. With simple modules there are also often more ways to implement a function allowing beneficial trade-offs between area and delay to be made.

On the other hand, using a module that is too simple can overburden the routing network. If a function that could be put into a few complex modules must instead be distributed among many simple modules, more connections must be made through the programmable routing network.

As a rule of thumb then, an FPGA should be as fine-grained as possible while maintaining good routability and routing delay for the given routing technology. The module should be chosen to implement a wide variety of functions efficiently, yet have minimum layout area and delay.

The approach used to select the Act1 modules was to employ macro usage statistics from masked gate array applications to evaluate candidate modules. This approach is

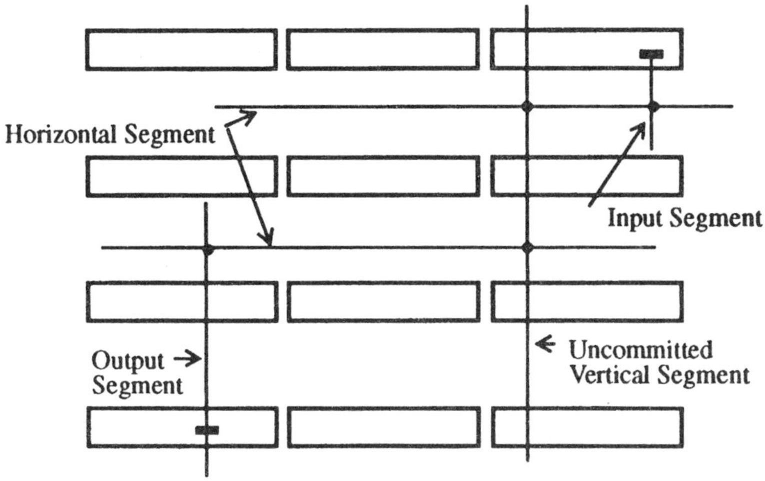

Figure 3.3.5. Routing Using LVTs in another Column

similar to that used to define the instruction set of a RISC microprocessor; choose a module that is most efficient for the most commonly used macros.

Given the parameters of antifuse routing, the effort focused on modules similar in complexity to a typical gate array macro. Statistics from designs targeted for implementation in a masked gate array were considered valid designs targeted at a candidate module. Only simple local optimizations (such as relocating inversions or combining a macro with a subsequent D flip-flop) influenced the module selection.

The Act1 family uses one general-purpose logic module [ElGamal1], shown in Figure 3.3.6 . Various macro functions (e.g. gates, flip-flops) can be implemented by applying input signals to the appropriate module input(s) and connecting other module inputs to logic 0 or 1. The module can implement any combinatorial function of two inputs, any function of three inputs (except the three-input NAND and exclusivity functions), many functions of four inputs, and other functions of up to eight inputs. Versions of AND gates and OR gates are implemented with all combinations of inverted and non-inverted inputs. In all, 702 distinct combinatorial macros are possible.

Any sequential macro can be configured from one or more modules using appropriate feedback routings. Over a range of designs, each module implements approximately 3.22 gates of logic; regardless of the ratio of combinational to sequential logic in the design.

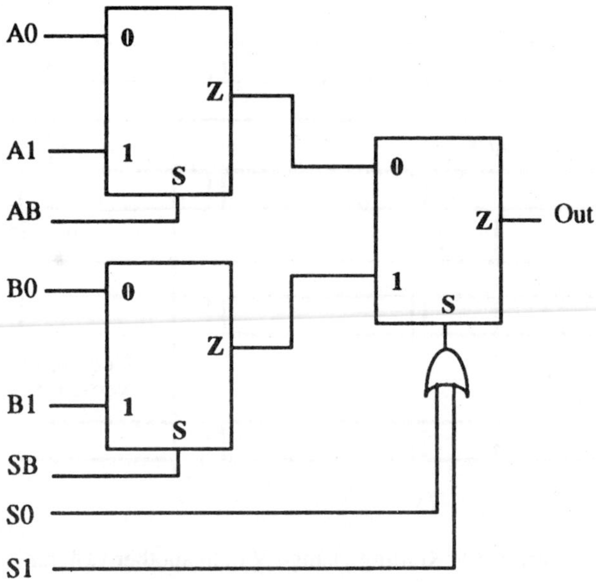

Figure 3.3.6. Act1 Logic Module

Each IO pad has an adjacent bidirectional buffer which connects to the array through an IO module. These modules are located in the outer columns and rows of the array next to the logic modules and interface to the routing channels in the same way as the logic modules. Each IO module has inputs for outgoing data and a three-state enable, and an output for incoming data which is driven from the pad. The IO module can be programmed to provide input, output, three-state or bidirectional capability.

The Act1 array is regular consisting of alternating rows of logic modules and routing channels, as shown in Figure 3.1.1 . The perimeter of the array includes IO modules. These modules connect the internal logic module signals to the pads. Each IO module has two inputs, data and enable, and an output. The data and enable signals are sent to the output buffer of the associated bonding pad, and the module's output comes from the input buffer of the pad. Thus the IO module can be configured to provide input, output, tristate, or bidirectional capability.

Clock signals present unusual requirements: they have high fanout and are required to have minimal delay and skew. Act1 devices meet the clock requirements with a dedicated clock network. Each network may be driven directly from an input pad for high speed. The signal passes through a buffer tree, and appears on a dedicated full-length horizontal track in each channel. Thus the network reaches every logic and IO module in the array.

Act2 Architecture

Statistical indications that most of the nets driving the data input of a flip-flop have no other destinations motivated the selection of two specialized modules for the Act2 family. The Combinatorial (*C-module*) shown in Figure 3.3.7 is similar to the Act1

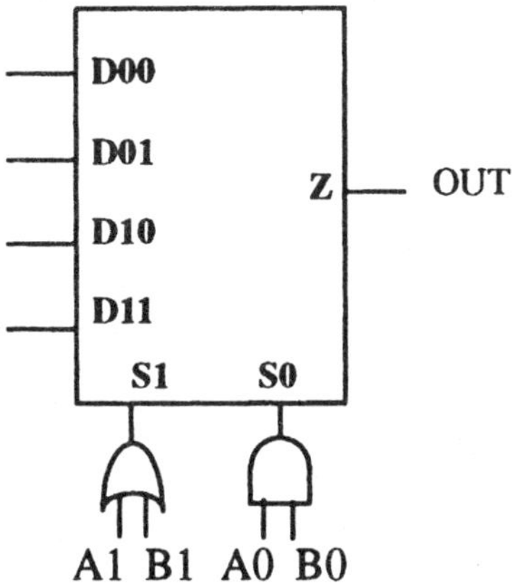

Figure 3.3.7. The Act2 Combinational Module (C-Module)

module. It was changed to better accommodate high fan-in combinational functions. It can implement 16 of the 20 four- input gates in the library, whereas the Act1 module implements eight, and some five-input AND and OR gates. The module modifications caused some loss in the ability to build sequential functions with C-modules. The module implements a total of 766 distinct combinational functions, including 13% more four-input and 12% more five-input macros than the Act1 module.

The Sequential module (*S-module*) consists of a front end equivalent to the C-Module followed by a sequential block built around two latches. Figure 3.3.8 gives a functional block diagram. The sequential block can be used as a rising- or falling-edge D flip-flop, or a transparent high or -low latch, by tying the C1 and C2 inputs to a clock signal, logical zero or logical one in various combinations. For example, tying C1 to 0 and clocking C2 implements a rising-edge D flip-flop. The block can also be set permanently transparent by tying C1 to 1 and C2 to 0, making the S-module

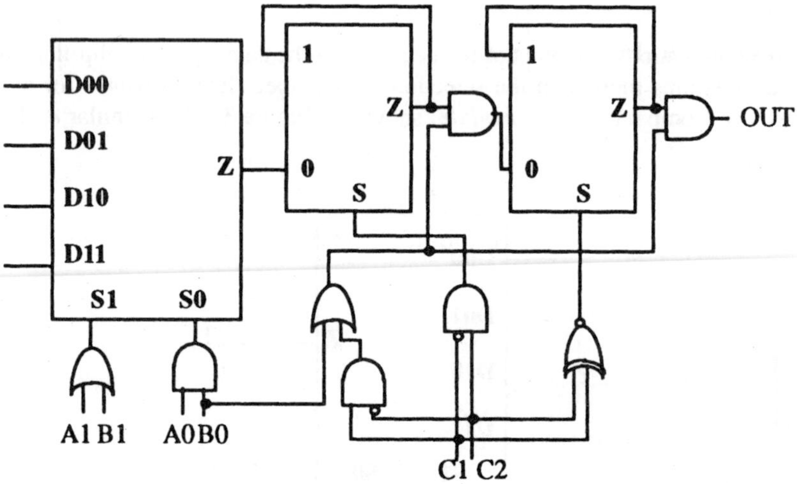

Figure 3.3.8. The Act2 Sequential Module (S-Module)

equivalent to a simple C-module with a small additional delay.

Figure 3.3.9 shows the functions that can be implemented using the S-module. Note that the latch with clear, a relatively rare macro, consumes part of the combinational logic to implement the clear; leaving a four-input function available in the C-Module. Toggle or enabled flip-flops can be made by using the logic in front of the D flip-flop. Other less commonly used flip-flops, such as JK or set/reset are not supported by the sequential block but may be configured from one or more C-Modules using external feedback connections as with the Act1 module.

A device with an equal mixture of C- and S-Modules provides a sufficient number of flip-flops for most designs with a margin to allow flexibility in placement. Over a range of designs, the Act2 modules provide about 1.4 to 2.0 times the logic capacity of Act1. As may be expected, the improvement is greatest for designs with a large proportion of flip-flops and wide gates.

Figure 3.3.10 shows a typical critical path in a state machine implemented with four C-modules and one S-module. The ability to fit a five-input gate in one C-module saves one routed net compared to the Act1 module implementation. The use of the S-module allows both the last combinatorial stage and the flip-flop to be combined into

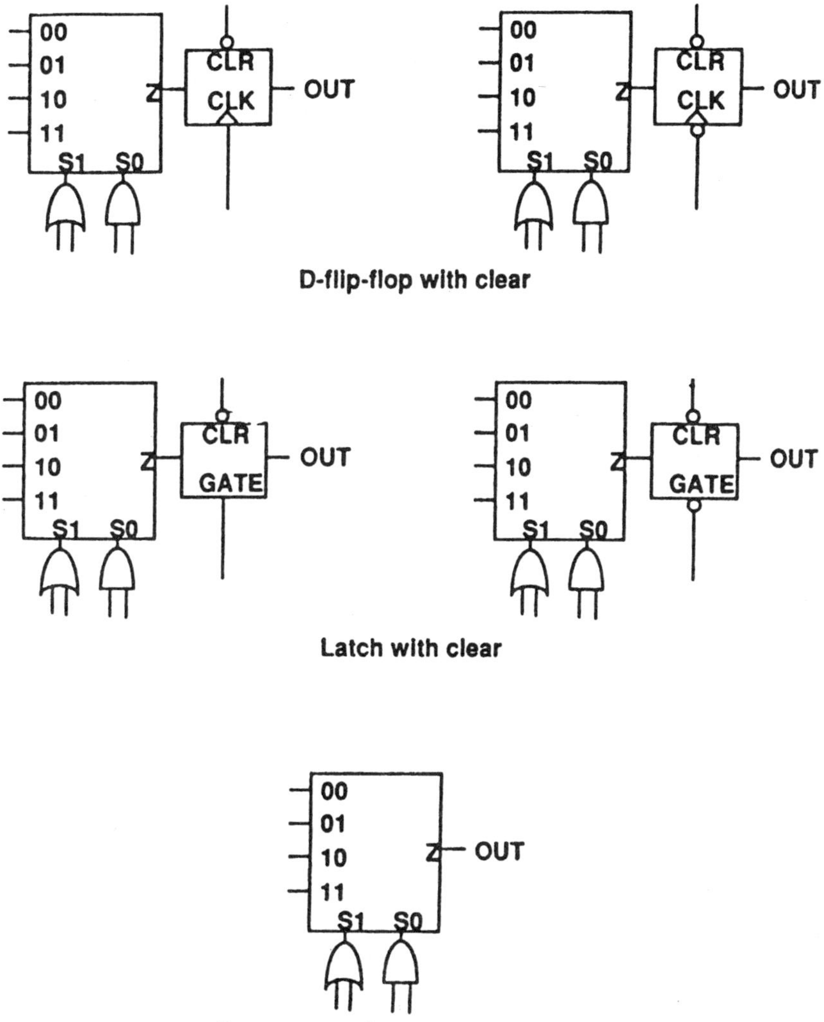

Figure 3.3.9. Act2 Sequential Module Configurations

a single module.

The choice of module also greatly influences the routability of designs. Because each input is accessible from only one of the two channels adjacent to a module, one might think that routability is worse relative to a conventional gate array cell, where a signal may enter from either channel. The impact of single channel accessibility, however, is not great because there is nearly always more than one way to implement a macro.

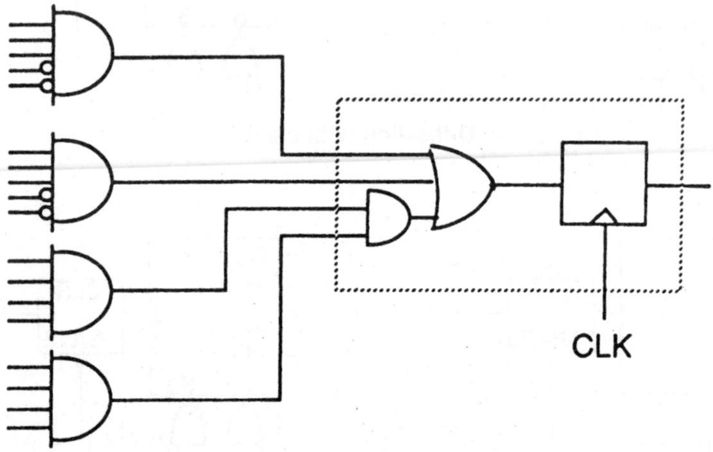

Figure 3.3.10. Act2 High Fan In Example

For an N-input macro there may be as many as 2^N different assignments of the input signals to the two channels. This corresponds to full double-entry symmetry. Even if not all of the 2^N assignments are possible, a sophisticated router can take advantage of implementation flexibility.

For the C-module, a given signal can be routed from either side of the module an average of 70 percent of the time. This number, weighted by macro usage, is quite sufficient for good routability. This figure is affected by both the module function and the assignment of a side to each module input. It is another important criterion in selecting a module function.

The module architecture of Act2 consists of alternating pairs of C-modules and S-modules. This provides the capability of implementing many macros that use pairs of combinational modules.

The Act2 family minimizes clock-to-output delay with a dedicated transparent-high latch in each pad, which can also be used as the slave stage of a flip-flop. The latch is controlled by a gate input to the IO module. If flow-through operation is desired, the gate is simply tied high to make the latch transparent. A dedicated transparent-low latch is provided on each input path. The polarities of the input and output latches are chosen so they can be combined with each other, or with other internal latches or flip-flops, to form a chain of rising-edge flip-flops.

Clock signals present unusual requirements: they have high fanout and are required to have minimal delay and skew. Act2 devices meet the clock requirements with two dedicated clock networks. Act2 also provides the option to drive the clock network from user-defined internal logic as well as an input pad.

To minimize the capacitive load on the clock network on Act2 devices, antifuses are only provided to connect the clock track to certain module inputs, specifically the clock inputs of the S and IO modules and a subset of the combinational inputs on all the logic modules. Skew may be further reduced by having the automatic placement program attempt to balance the loading on each branch of the distribution tree.

Like other inputs, the clock inputs to the S-module, may be connected to the normal horizontal segments. This accommodates designs with several local asynchronous clocks. As with any technology, using asynchronous clocks with ACT FPGAs can cause race conditions and the designer must be alert to avoid them.

Act3 Architecture

The Act3 architecture is actually a twist on Act2, with key features added that optimizes the part for speed and flexibility. The Act3 logic module is very similar to Act2, in fact only two changes were made to the logic module, both targeted to increase flexibility and sequential system speed. The first change was to add a *clear* input to the sequential module, thus making the combinational portion of the sequential module completely independent from the sequential logic. There is a direct area cost to adding an input, but this addition provides a more uniform module for synthesis, and therefore increases the Actel flexibility as shown in Figure 3.3.11 . A user of the Actel array benefits greatly by being able to map more logic functions into the combinational portion of an S-Module. This aspect of the S-Module is called *combinable,* and is significantly better in Act3 over Act2. Figure 3.3.12 is a diagram of the S-Module, a C-Module corresponds to the Combinational block.

The other significant change in the Logic Module Architecture is the addition of a fast clock. This clock is not routed, and is significantly faster than its routed counterparts. The clock select, in Figure 3.3.12 , allows the selection between the high speed clock network, and the more flexible routed clocks.

Figure 3.3.11 illustrates the additional combinable logic functions available in the Act3 sequential logic module, as compared to Act2.

The other significant change in Act3 is the IO architecture. The Act3 is IO module rich, with flip-flops available on both the input and output paths. Figure 3.3.13 shows the IO Module. The IO module incorporates a separate fast IO clock. Synchronous enable signal for the flip-flops are also available as well as an output register feedback signal. This IO architecture is targeted for speed, and supports a

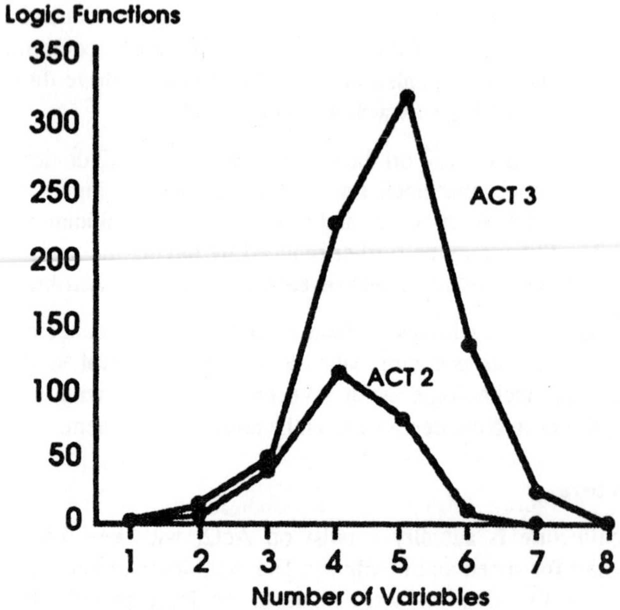

Figure 3.3.11. S-Module Combinatorial Logic Functions

significantly faster clock-to-Q, currently measured at 10 nanoseconds.

Programming and Testing

One of the great puzzles in the development of antifuse FPGAs was to find an efficient way to uniquely address each of the two-terminal programming devices. Diodes in series with each antifuse would allow unique addressing, but block signal flow. Early schemes required the use of an individual control line for each routing track, doubling the number of lines required in each direction [Graham]. Methods for programming and testing antifuse FPGAs that use only a few control lines for an entire channel were developed for ACT devices [ElGamal] [Ahrens]. Although a full description is beyond the scope of this text, the following explanation conveys the basic concepts.

Consider an array of antifuses at the intersection of some horizontal and vertical segments, as shown in Figure 3.3.14 . An antifuse is programmed by applying a programming voltage *Vpp* across it. This is done by precharging all segments to an intermediate voltage of about *Vpp/2*. Then a selected vertical segment is grounded and a selected horizontal segment is driven to Vpp. Other segments are left floating at Vpp/2. Only the single antifuse at the intersection of the selected segments sees the

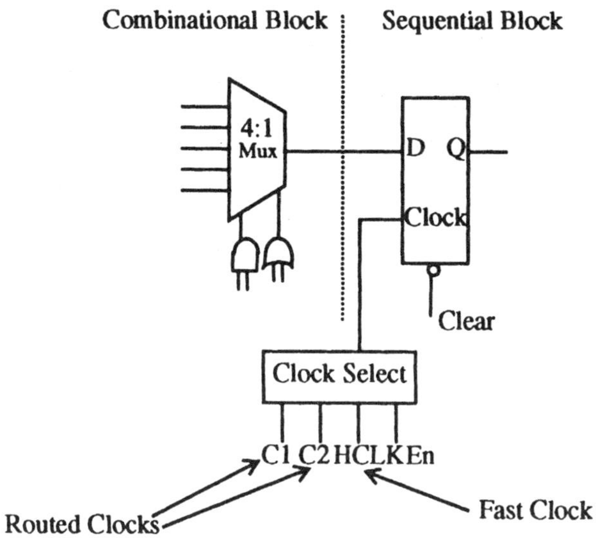

Figure 3.3.12. The Act3 Sequential Module

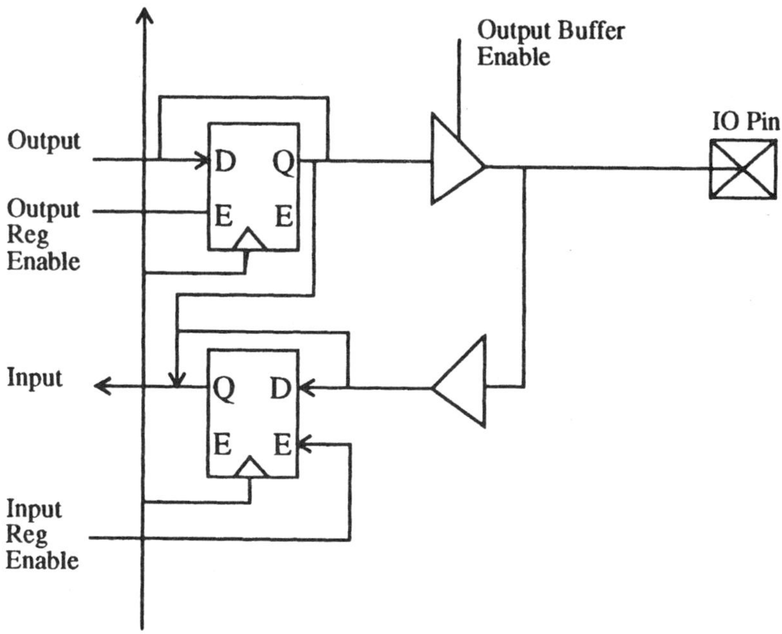

Figure 3.3.13. Act3 IO Module

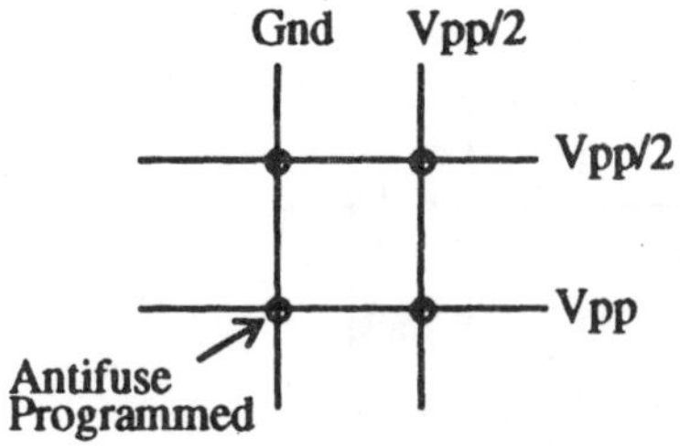

Figure 3.3.14. Programming an Antifuse

full Vpp.

Only the unique antifuse with the full Vpp potential across it is programmed. Figure 3.3.15 illustrates the *series pass transistor addressing* method. The structure is similar to that of Figure 3.3.3 with the modules removed for clarity. Each pair of adjacent segments in the same horizontal track is connected by a pass transistor. In tracks containing uncommitted segments, there is an antifuse present between the two segments, and so the transistor is connected in parallel with the antifuse.These transistors are used only for programming and testing and are shut off during normal operation of the programmed part. The transistors in each row of modules are gated by a control line, marked as *Row Control Line* in the figure, and the transistors in each column of modules are gated by a control line, also marked in the figure. These lines are driven, in turn, by programming logic present at the array periphery. Note that only one control line per row or column is required, regardless of the number of tracks.

Figure 3.3.16 illustrate the circuitry used to program an antifuse at the intersection of a horizontal and vertical segment. The vertical track containing the antifuse is driven from one end to ground by the programming logic present at the array periphery. The pass transistors in all rows between the driving periphery and the antifuse are turned on. The horizontal track is driven to Vpp as shown in the figure. A fuse between two adjacent segments in the same track is programmed in a similar manner. All columns of pass transistors except the one bypassing the antifuse are turned on, and the track is grounded at one end and driven to Vpp at the opposite end.

Figure 3.3.17 illustrates *direct addressing* circuitry, which is some cases offers a favorable alternative to series pass transistors. Column voltage supply lines run vertically through the array. Each segment along a track may be connected through its own addressing transistor to a supply line. These transistors are gated by a horizontal

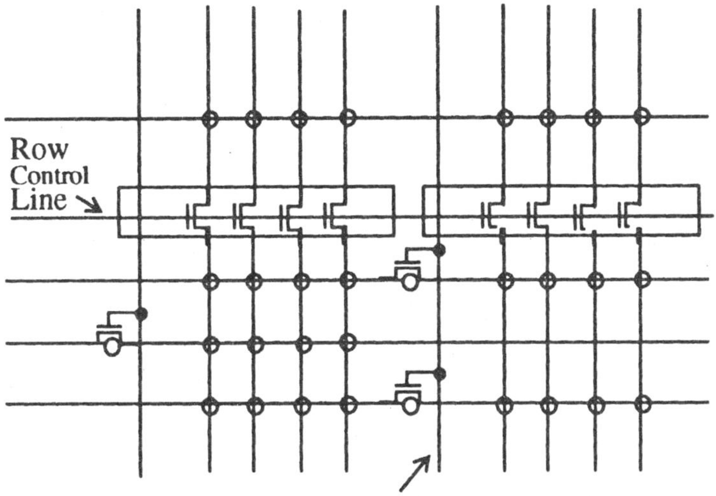

Figure 3.3.15. Antifuse NetWork

select line. Each select line serves all the segments in a group of one or more adjacent tracks in the same channel. Activating one vertical supply line and one horizontal select line uniquely addresses a particular horizontal segment. The number of segments in a channel that can be addressed is limited by the number of supply lines times the number of group select lines. It follows from this that the ratio of tracks to select lines is at most the average segment length. An antifuse between two adjacent segments may be programmed by activating the control line for both segments and two supply lines, one for each segment. Since only one transistor lies between a segment and the supply, the programming current is independent of the position of the segment and the segmentation of its track. Thus the direct address method is most efficient for irregular channels with long segments.

Act1 devices used only the pass transistor method. Act2 and Act3 devices use the pass transistor method for the vertical tracks, which contain many short input segments, and the direct address method for the horizontal tracks.

In either scheme, some care is required to assure that a unique antifuse is addressed once other antifuses have already been programmed. Figure 3.3.18 provides an

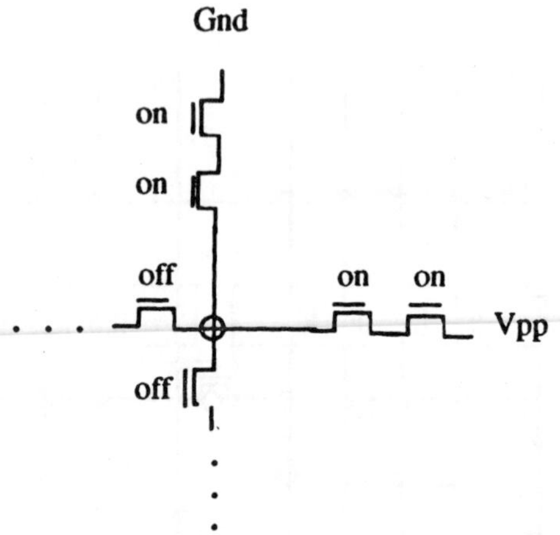

Figure 3.3.16. Antifuse Programming Example

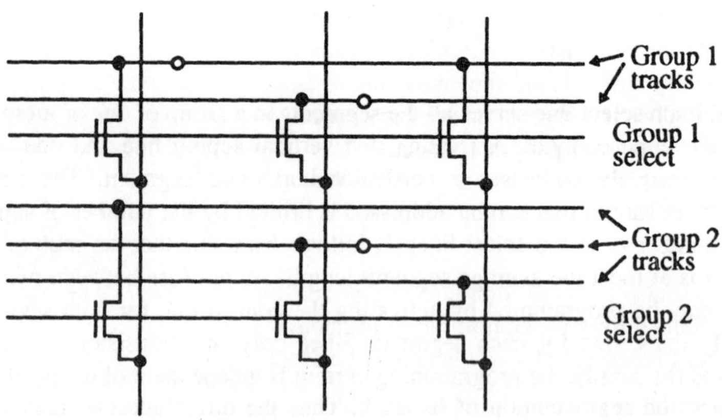

Figure 3.3.17.Direct Addressing Of Fuses

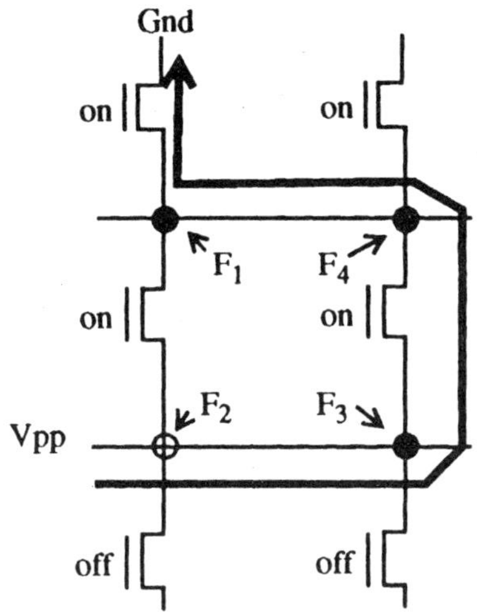

Figure 3.3.18. Programming Sneak Path

example of how improper addressing allows programming current to divert along a *sneak path.* The previously programmed antifuses F_3 and F_4, cause the programming of antifuse F_1 instead of the intended antifuse F_2. Unlike PROMs, FPGA programming does not involve an arbitrary pattern of antifuses. For example, it is not necessary to program a pattern connecting two outputs together since this does not form a useful net. For the relevant patterns, it may be shown that there is always an order in which the antifuses may be programmed with no chance of a sneak path occurring.

Special care is also required to protect the module circuitry from the voltages present on the segments during programming. Transistors that are in contact with the routing segments must be designed to withstand to the programming voltage.

Device testing takes place in three phases: before, during and after programming. Pre-programming tests check for shorted or open segments, shorted or even relatively low resistance antifuses, and proper module and IO operation. Continuity of the segments is easily tested by turning on all pass transistors and using the peripheral circuits to drive the segment at one end of a track and read the segment at the other end. Testing for the absence of shorts between segments in adjacent tracks is done similarly by applying a pattern of alternating zeros and ones. Weak antifuses may be screened out by applying the proper stress voltage (higher than normal operating voltage but lower than Vpp) across groups of antifuses in parallel using the programming circuits. Breakdown of an antifuse is detected by passage of excessive current through it.

To verify the functionality of the modules we need to apply test vectors to their inputs and read their outputs. A vector may be applied simultaneously to an entire row of modules by turning on all vertical pass transistors except those in the row being tested. Data are applied to the inputs in the channel above the row from the top periphery and to the inputs in the channel below the row from the bottom. A simple row-select and column-sense scheme conveys the output of each module in turn to an output pad for monitoring.

Proper closure of the programmed antifuses is verified during programming by sensing the passage of programming current. A complete test for unintended connections between any two segments can be done after programming using the programming circuitry to precharge, drive, and read the segments. This is true despite the fact that it is no longer possible to uniquely address each individual antifuse once programming commences. The reason is that detection of shorts, but not the location, may be accomplished simultaneously for many antifuses in parallel. Taken together, the tests described insure correct and complete functioning of the programmed part.

Capacity

Determining the capacity of FPGAs is a complex task with many considerations. ACT device capacities are specified in terms of masked device gates. Masked gate arrays are specified to contain a number of two-input NAND gates. Utilization of the gates is a measure of the percentage of the total number of gates on the device that a typical design may use before running out of routing resources.

In an FPGA there are two utilization questions: how many of the modules can be placed and routed, and how many gates are obtained per module. ACT devices have a small logic module with the low granularity of a gate array as well as sufficient routing resources to achieve high module utilization. The module utilization of ACT devices is guaranteed to be 85% or greater with many designs achieving utilizations over 95%.

Both Act1 and Act2 devices have been tested for capacity using four well-known benchmarks. The results appear in Table 5 . Customer designs have also been used to analyze Act1 module gate utilizations. On average, Act1 modules implement 3.22 gates per module for all types of logic. The mixture of gates to flip-flops in a design does not affect the gate utilization of Act1 devices.

Table 5: Benchmark Capacity Results

Device	Benchmark	Gates	Instances	Total Gates
A1020	Data Path	157	12	1884
	Timer/ Counter	248	6	1488
	State Machine	153	9	1377
	Arithmetic	295	6	1770
Average				1630
A1280	Data Path	157	54	8478
	Timer/ Counter	248	26	6448
	State Machine	153	37	5661
	Arithmetic	295	18	5310
Average				6474

Table 6 summarizes results for a variety of designs in the A1280 and A1240 FPGAs. The designs were placed and routed automatically with no manual intervention. For this reason, the table includes the number of routed pins per logic module as a measure of design complexity. Nearly all designs with module utilization under 85%, most designs with utilization under 95%, and many with utilization up to 100%, route completely.

Table 6: Capacity Benchmarks

1280 FPGA 1232 Logic Modules	Logic Module Utilization	Pins Per Logic Module
Design done in an 8K TI Gate Array	99.5%	4.28
32 Bit Data Path, 16X16 Mult., State Machine	99.4	4.30
2901 ALU (X4)	98.1	4.57
DMA Controller (X3)	97.1	3.99
Asynchronous Serial ECC	97.0	4.78
Pipelined Fixed Point Mult, Div, Sqrt	94.5	3.37
State Machine, Mult/Add, Datapath, Counter	92.7	4.34
Color CRT Controller (X3)	87.3	3.83
32 Bit Datapath with Sum, Compare (X3)	86.8	5.25
40 Bit FLoating Point Adder/Subtractor	86.7	4.33
1240 FPGA 649 Logic Modules	**Logic Module Utilization**	**Pins Per Logic Module**
16 Bit Datapath, 16X16 Mult., State Machine	98.1	4.86
2901 (X2)	93.2	4.68
DRAM, DMA and SCSI Controllers, UART	92.6	4.73

Performance

Like capacity, performance measurements of FPGAs is problematical. Specifications such as maximum clock frequency or module propagation delay are insufficient indicators for estimating the performance of a given design. Like masked devices, the propagation delay of FPGAs is routing and fanout dependent. The ACT databook gives numbers for propagation delay based on device characterization for various fanouts.

Figure 3.3.19 shows the cumulative distribution for the routing delay to each input pin for various ranges of net fanout. The histogram includes all nets in the design. Pins on nets identified by the user as speed critical would generally be among the fastest 30%, and nearly always among the fastest 90%. The routing delay distributions are reasonably tight, especially for low fanouts and critical nets.

Table 7 shows results from tests on some typical circuits. The table shows the measured performance for the Act2 1240-2 and the Act3 1425, as of June 1993.

Table 7: Performance Benchmarks

Example	Speed (MHz)
Act2 1240-2	
16 bit counter	58
16 bit accumulator	36
state machine	39
Act3 1425	
16 bit counter	68
16 bit accumulator	40
state machine	46

The Act3 architecture is specifically targeted for speed. The measured on-chip performance is 125 MHz with 10 ns clock-to-out. Worst case performance benchmarks show 40 MHz 24-bit accumulators, 75 MHz 24-bit loadable counters, 125 MHz 24-bit pre-scaled loadable counters, 125 MHz shift registers, and 80 Mhz chip-to-chip speeds.

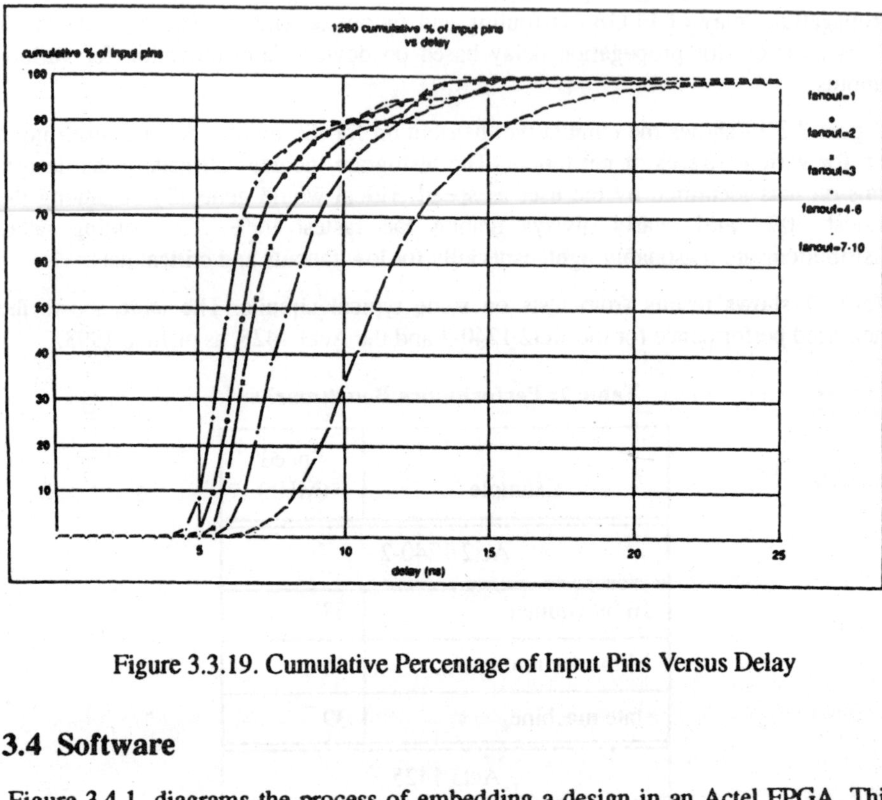

Figure 3.3.19. Cumulative Percentage of Input Pins Versus Delay

3.4 Software

Figure 3.4.1 diagrams the process of embedding a design in an Actel FPGA. This process begins with capture of the design in a computer readable format. Currently most users enter their designs as schematics built of macros from a library. Any of several standard schematic capture programs can be used. Since the Act1 and C-modules have the same complexity as a typical gate array library macro, the problem of selecting which groups of macros should share one module is avoided; each macro is typically assigned its own module. The larger capabilities of the S-module are handled by software that automatically combines a flip-flop and a preceding combinational macro into one S-module where possible. This process is transparent to the user and does not require modifying the schematic.

Another increasingly popular alternative is to enter designs in terms of Boolean equations, state machine descriptions, or functional (rather than structural) schematics. Different portions of a design can be described in different ways, compiled separately and the results merged according to a top-level hierarchical

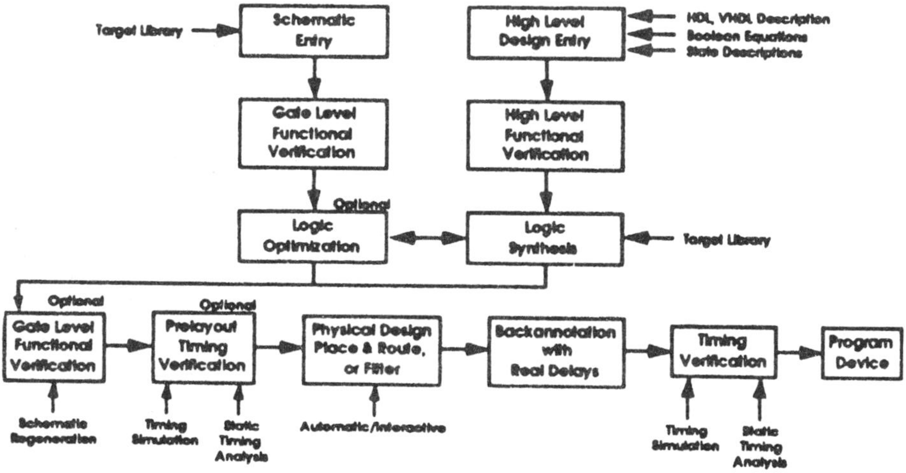

Figure 3.4.1. FPGA Design Flow

schematic. Various industry standard formats are supported.

High-level synthesis tools such as MIS-II [Brayton] or Synopsys [Rudell] can be used with Actel FPGAs by providing a suitable library. These tools can either encompass only the standard schematic library macros, or include the larger set of all 700 functions embeddable in the typical Actel module. Recent research has investigated other more efficient ways to allow synthesis tools access to the full flexibility of the module. A method for rapid searching of libraries using Boolean matching is given in [Mailhot]. Other approaches take advantage of the multiplexor structure of the module, using either binary decision diagrams [Murgai] or *if-then-else DAGs* [Karplus].

Several guidelines are suggested for reliable logic design with FPGAs. The general goal is to make proper circuit function independent of shifts in timing from one part to the next.

- Use synchronous logic design where possible.
- Avoid gated clocks, using enabled flip-flops instead.

- Avoid race conditions.
- Limit fanout by buffering.

These guidelines are the same as those for users of mask programmed gate arrays [LSI]. In addition, designs for Actel FPGAs must use multiplexers rather than tristate drivers since internal tristate is not supported.

Once the design has been entered it may be simulated either functionally or using pre-layout delay estimates before entering the ALS system. ALS provides for the selection of a device, package before the netlist is checked by a validation program for problems such as undriven nets, outputs shorted together, excessive fanout, etc.

A good pin assignment is key since it can impact the design's use of routing resources. Pin assignment may be done manually, or by the software, which bases its assignments on an analysis of the design.

The next step is to map the netlist into the Actel architecture. The placement problem, or selecting a module for each macro, is similar to that of a conventional gate array. Since macro placement effects all aspects of the design, including its speed and the number of gates used, there are a number of different placement options available. Automatic placement used algorithms similar to gate array programs, with a few additional considerations such as fixed power and programming pins. Incremental placement, as shown in Figure 3.4.2 , takes a netlist that has been modified slightly and updates the current placement slightly. Incremental placement is useful because it maintains similar net delays for subsequent placement runs.

Since the key is ensuring that critical nets meet speed requirements, another desirable alternative is Timing Driven Placement. Such a program performs placement based on timing constraints on speed critical paths. Current research suggests that definite improvements in routing delays can be made by applying these techniques.

Routing, described at length in "Principles of Programmable Routing" in section 3.3, is the next step. Routing congestion within each horizontal or vertical channel must be limited. Whenever possible, the dedicated output segments should be used in preference to the uncommitted vertical segments, especially for nets identified by the user as speed critical. Routing of the segmented channels is done as previously described. On Act2 devices, macros must be placed in modules of the appropriate type (C or S). Time for complete placement and routing of an 8000-gate A1280 FPGA is about 30-40 minutes on a 486.

Once placement and routing are completed, the propagation delay to each input pin is calculated based on device characterization data. The calculation accounts for the module internal delay as well as the delay through each RC stage in the routed net. The estimates can approach the accuracy of a SPICE circuit simulation. The delays may be back- annotated to a simulator, or checked with a static timing analyzer. In either case, timing data may cause an iteration in the design. At some point the design

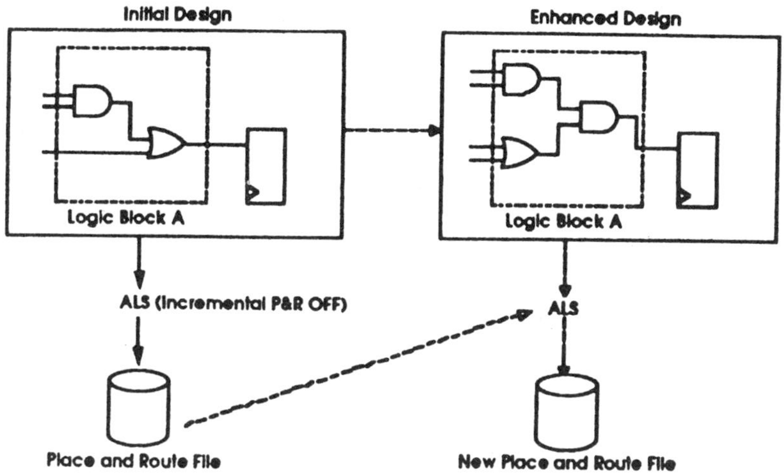

Figure 3.4.2. Incremental Placement Design Flow

meets whatever criteria the user is targeting, and it is time to program the FPGA part.

A list of antifuses to be programmed is generated and downloaded to a programming station which also holds the FPGA. Programming time is about 5-10 minutes, depending on the size of the FPGA and the design. The Actel Activator2 station allows up to four chips to be programmed simultaneously. The standard tests applied before, during and after programming, described in section 3.3 assure correct implementation of the programmed part. No design-specific test vectors are required from the user.

The programming system can access the test circuitry inside the programmed FPGA allowing it to sense the value of any pair of selected module outputs and present them on an external pin in real time. This unique probe feature, called the debugger, can be used as a debugging tool for the device while it is operating at speed in the system. The probe provides a useful back-up when simulation is difficult or impossible, such as for highly asynchronous designs or when the design must accommodate a poorly documented interface. Debugger software is also provided to enable the programmer to be used as a functional tester, presenting stimuli at the FPGA's pins and reading data back via the probe and the pins. Figure 3.4.3 illustrates the Actel design system.

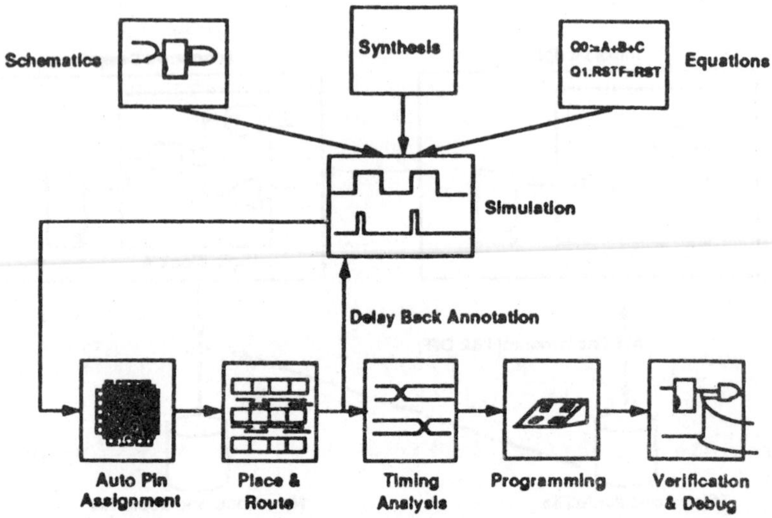

Figure 3.4.3. The Actel Design System.

3.5 The Future

How will the speed, density and cost of antifuse FPGAs evolve? Antifuse FPGAs will directly benefit from anticipated improvements in the underlying CMOS technology. In addition, it is likely that further improvements will be made in the antifuse itself. Changes that reduce the resistance, capacitance, or programming voltage would be beneficial and could significantly benefit architecture possibilities. On the architecture front, better logic modules and improved methods of designing segmented routing channels may be developed.

Antifuse technology provide the opportunity of developing sea of module (SOM) architectures, unlike other FPGA technologies. Just as gate array architectures placed the routing channels over the modules, the antifuse based channeled architecture has the same potential. A sea of module architecture can reduce the die area by up to half of the array size. Current process technologies have the promise of being able to place the antifuse on top of the logic modules.

3.6 Design Applications

Designing with ACT1 and ACT2 FPGAs

The advantages of using FPGAs over discreet devices are well known. In order to maximize the benefits of high integration and low power, it is important for the designer to understand how applications are mapped into the architecture, and trade-offs in applications. It is up to the designer to understand how to select components from the library so as to take best advantage of the logic modules. While it is not required to understand the architecture of Actel FPGAs to implement designs with them, there are architectural considerations that would aid designers if they kept them in mind when entering a design. Some considerations apply to all Actel devices, others are architecture specific. The differences between Act1, Act2 and Act3 devices also suggest some differences in the design considerations for using each of the families of devices. The applications in this section use Act1 and Act2. The Act3 applications are similar to Act2 only faster.

In order to take best advantage of the logic integration capability of an FPGA, designs that target such devices should implement functions using the fewest number of logic modules possible. The issue of capacity is closely related to that of efficiency. It also relates to performance. Sometimes the fastest design is also the smallest. In other cases using parallel logic increases the module count, but improves performances by reducing the number of logic levels.

An efficient implementation is one that uses the fewest number of logic modules to implement as many equivalent gates as possible. An inverter would be a very inefficient use of an Actel module while a single-bubbled input AND-EXOR (equivalent to 4.5 two-input NAND gates) would be very efficient.

Inverters or other library cells that are inefficient should be avoided. In most cases bubbled inputs on macros are available to implement inversions at no cost in modules. Two-level cells (e.g. AND-OR) are advantageous from both a capacity and performance perspective. Such cells compress logic to reduce delays and increase the number of gates per module.

Performance

The performance of logic is variable, so it is important to optimize designs. There are four criteria that influence performance of logic in FPGA designs. They are listed below in descending order of importance:

- **Levels of Logic** The fewer number of combinatorial logic levels between flip-flops the faster the logic.
- **Fan-out** Propagation delays in FPGAs are sensitive to fan-out. Limiting fan-out on individual nets improves performance.

- **Fan-in** Measures the number of nets connected to a logic module's inputs. Library functions with heavy fan-in efficiently utilize the logic of the module and aren't a problem when used sparingly. Too many high fan-in macros, however, can congest routing and reduce performance.
- **Number of modules** Fewer logic modules allow them to be placed closer to each other. Shorter distances between modules speeds the connection paths.

Chip Level Design Considerations

While each of the above considerations is important in itself, it must be remembered that they are interrelated and an improvement in one may cause a degradation in another. For example, limiting the number of logic levels tends to increase fan-in. If the design has a significant number of high fan-in macros, use of additional high fan-in macros in a counter may cause routing congestion. A balance must be found among them so than none becomes a drag on performance. The optimal solution to a design may only be found after some iterations that adjust for such things as fan-out and criticality. The ALS tools such as the *Validator* and automatic Place and Route can rapidly provide answers to questions about design capacity and routability.

Act1

The multiplexor-based logic module is highly efficient in multiplexor implementations allowing versions of both 2:1 and 4:1 to be implemented in one module. The Act1 library contains all bubbled input permutations of AND, OR, NAND, and NOR gates. All two-input and virtually all three-input gates are configured from a single module.

Eight of the twenty four-input gates offered require one module while the rest use two modules. Using the single module gates saves module resources and may be done by matching bubbled driver outputs with bubbled gate inputs as shown in Figure 3.6.1 . When a two-module gate is required, the delay penalty is not as great as two levels of gating because the two modules of the macro must be placed adjacent to one another.

Many highly efficient multi-gate macros are available in the Act1 library. They include AND-OR, OR-AND, and AND-EXOR functions. Designers should become familiar with the library so they can make the best use logic resources on the device.

A single Act1 module can be used as a latch, while flip-flops require two modules. Latches should be used whenever the design allows it. When flip-flops are required you should use those that the highest number of gates for the modules. There are several enabled and multiplexed flip-flops available in the library.

A flip-flop enable can be used in place of an AND gate driving the data input. The two-input multiplexor on a flip-flop can be used as a multiplexor or it may be used as a gate by tying one of the multiplexor inputs to a rail. The possible configurations are shown in Figure 3.6.2 .

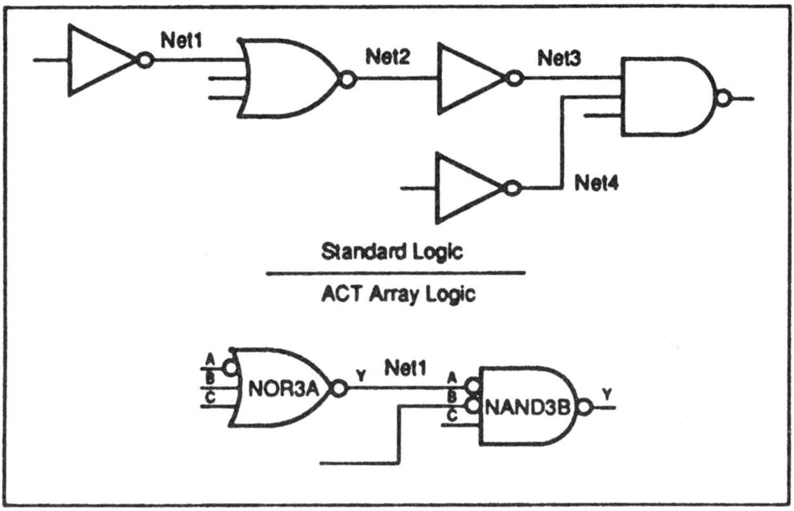

Figure 3.6.1. Reducing Logic Complexity

Act2

Before proceeding to a detailed description of design considerations it would be useful to review some fundamentals of the Act2 architecture. The Act2 architecture features two types of modules. Combinatorial modules (C-modules) are used to implement any combinatorial function in the Act2 library. Sequential modules (S-modules) can be used for either sequential functions (e.g. flip-flops) or combinatorial functions or both. When the S-module is used to implement both a sequential and a combinatorial function (e.g. a gate followed by a flip-flop) it is being used in the most efficient way. The ALS software will automatically combine a flip-flop and a preceding combinational macro into one S module where possible and perform other simple local optimizations such as eliminating unused modules in soft macros. Whenever a flip-flop and the combinatorial function driving it are combinable, and the net between them drives no other macros, then both will be combined into an S-module.

The module types exist in roughly equal numbers on Act2 devices and the place and route software will automatically select the appropriate module for each library component in the schematic.

The Act2 library is compatible with the Act1 library. Because the Act2 C-module is more powerful than Act1 modules it can implement more functions in a single

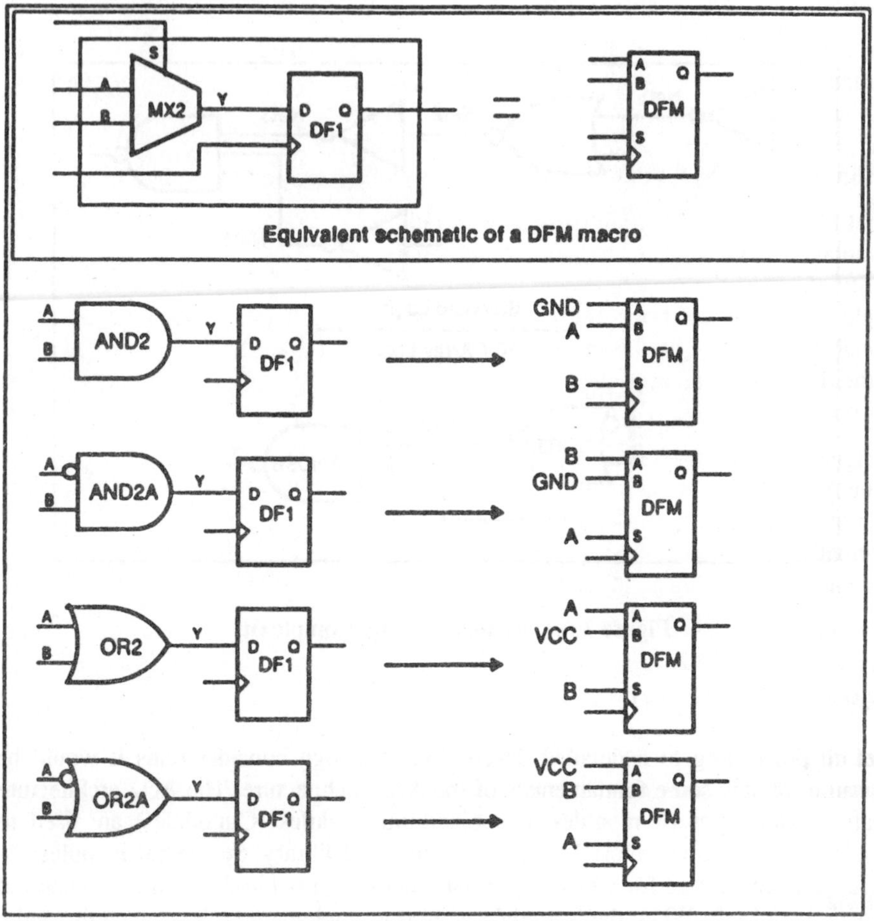

Figure 3.6.2. Compacting Logic

module. All the three-input gates and all but four of the four-input gates are single-module macros. There are also four five-input gates that require only one module.

Because about half of the modules on Act2 devices contain dedicated flip-flops, the function is inexpensive in terms of resource requirements compared to Act1. The abundance of flip-flops coupled with their high performance means that they are more attractive as a sequential element than latches.

Flip-flops may be implemented using an S-module, an S-module and a C-module, or (as in Act1) two C-modules depending on functionality. The ALS documentation describes how each function is done and whether it is combinable.

The Act2 device has a built-in latch on each IO pad. The latches can be used as storage element for inputs, outputs, or bidirectional IOs. An IO latch may be used with an internal latch to make an IO flip-flop.

Designing with ACT FPGAs: A TTL Perspective

Actel FPGAs offer many advantages over traditional technologies such as TTL. The advantages include greater reliability and reduced board space and power from the ability of FPGAs to integrate large amounts of logic into one device. For example a single A1280 FPGA holds the equivalent of 165 MSI TTL devices (assuming 40 gates/MSI device). That means not only a smaller PCB, but a simpler one since most of the designs connections are made inside the FPGA by the 100% automatic place and route software.

Designers who are used to using TTL components may see some of the advantages of using Actel FPGAs in their designs and not realize how easy it is to begin using them. It's not necessary to know anything about the architecture of the FPGA. In fact, the schematic entry and simulation process is the same as it was with TTL. The placement and routing are done by software tools that are analogous to PCB software.

Actel provides a library with the system for popular schematic design tools. The library contains both hard macros and soft macros. Hard macros are similar to SSI components. They form the basic functional building blocks such as gates and flip-flops. Many Actel hard macros are identical in function to TTL parts though they have different names. The Actel databook contains a cross reference guide showing the names of hard macro components that match functionally to TTL.

Soft macros are more complex functions that are built from some number of hard macros. They include counters, decoders, and adders of various sizes. Some of the soft macros in the library have identical functions to MSI TTL parts. These may be identified by their name beginning with TA instead of 74. The rest of the name matches that of the TTL name. Other soft macros offer generic logic functions.

All soft macros are easily copied and modified, and there is no limit to the number of custom soft macros you may add to the library. Should you need a TTL function for which there is no near equivalent in the library it is easy to build from scratch. Simply copy the schematic as shown in the TTL databook using components such as gates and flip-flops from the Actel hard macro library.

As you gain familiarity with the Actel hard macros you will find instances where you can do a more efficient design than that found in the TTL databook. For example, if the book shows an AND gate driving an OR gate you may find a single hard macro containing both the AND and OR functions. Using such multifunction macros is very efficient because you get more and faster logic from the macro. Compare the

74AS161 counter schematic from a TTL databook with the TA161 from the Actel library in Figure 3.6.3 . The Act1 version of the function uses only 18 modules, or 3.3% of an A1020.

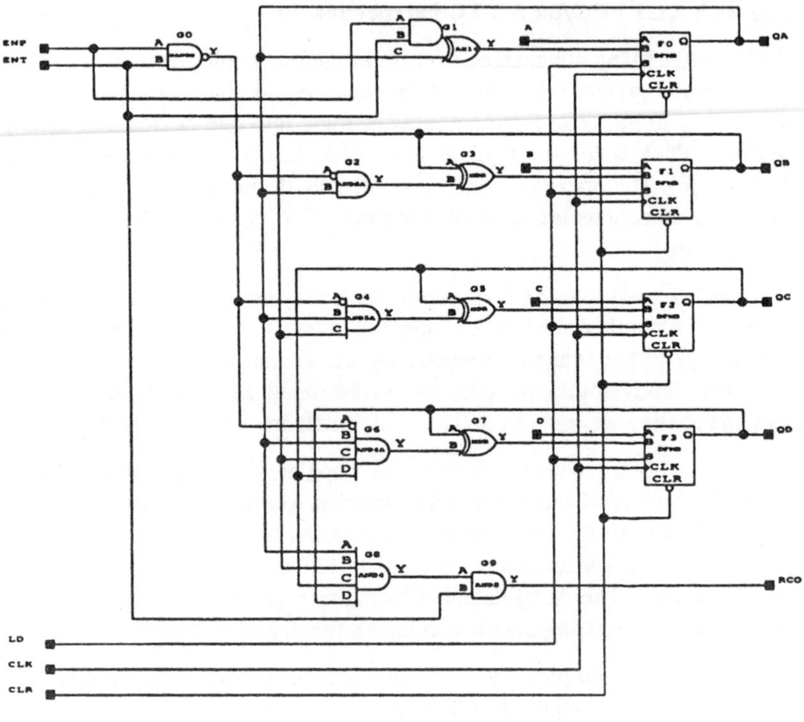

Figure 3.6.3. TA161 from the Actel Library

Many TTL parts have three-state outputs allowing them to be connected to a common bus. Three-state functions don't work well with ASICs or FPGAs because they tend to be slow and inefficient.

You don't have to give up using busses in your designs. Simply implement them using multiplexors as shown in Figure 3.6.4 . Multiplexors are very efficient on Actel parts. For example, you can create an eight-bit bus with four possible drivers using less than 3% of an A1010.

If you use a soft macro, but don't need all the outputs, you don't need to modify the macro. The Actel software contains a program called the gobbler which will

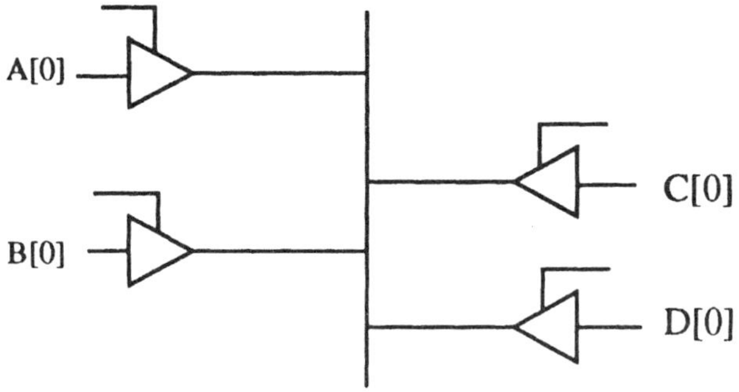

Discrete Technology Implementation

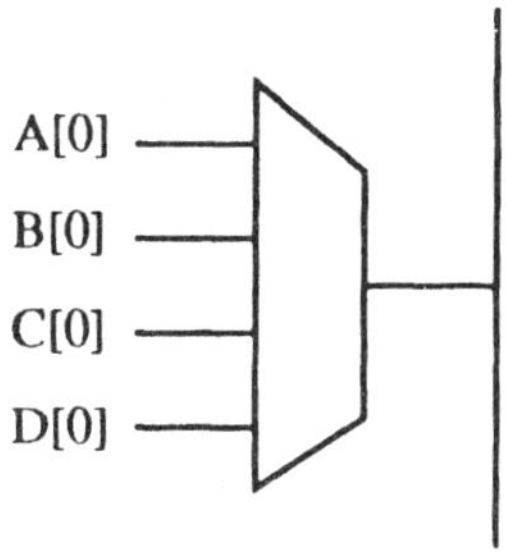

Actel FPGA Implementation

Figure 3.6.4. Least Significant Bit of a Bus with Four Possible Drivers

automatically eliminate any unused modules before the design is placed and routed.

If you use a soft macro or hard macro that has inputs you don't need the situation is different. The software won't allow inputs to be left unconnected, so some designers simply tie unused inputs to VCC or GND. That is permitted, but a better solution would be to select a hard macro that only has inputs you need or modify the soft macro to eliminate the unused function.

For example, the TA161 counter has a load function and four data inputs. Rather than tieing off those pins, a better solution would be to make a copy of the counter and modify it as shown in Figure 3.6.5 That will allow the wiring resources on the chip to be used for things other than power and ground connections.

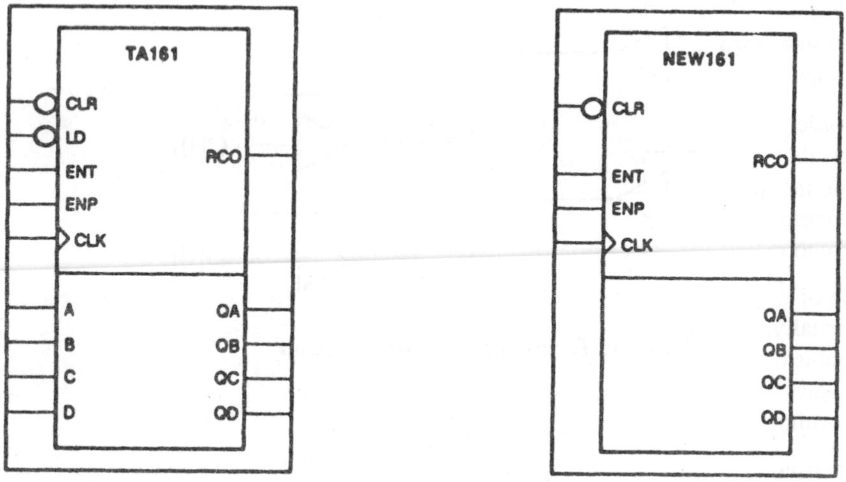

Figure 3.6.5. Old and New Versions of the TA161 Counter

Migrating PLD Designs to FPGAs

Designers who have been using discreet technology want a seamless method to convert old designs to Field Programmable Gate Arrays. They also want to have a way to enter new designs using tools and techniques they are familiar with. Tools are available that allow designers to take advantage of the logic integration capabilities of FPGAs (with the concomitant benefits of low power and high reliability) while retaining familiar design methods. The tools allow for migrations of PLD designs to be done rapidly using PLD logic description files and without drawing a schematic rendition of the design.

While the PLD migration tools can insure functional compatibility with the PLD version of the logic, designers rightly wonder about such issues as performance, efficient use of FPGA resources, and integration of the PLD logic with other functions now on the FPGA.

We'll explore some of the issues and provide a design flow for migrating PLD designs to FPGAs using the Actel synthesis tool ACTMAP which accepts a PALASM file as input and produces a netlist of Actel library components. This particular synthesis example is used for illustration only, Actel supports many different synthesis tools, and the results of each will vary, but the basic approach to synthesis is similar.

Unlike PLDs, the performance of a function implemented in an FPGA can vary.

Variations come from differences in fan-out and placement of logic modules. While it is not possible to know precisely what the performance of a function such as a state machine will be prior to placing and routing a design, pre-route estimates are reasonably accurate. Performance requirements for the PLD design on an FPGA can be stipulated to synthesis tools.

In order to take advantage of the logic integration capability of an FPGA, synthesis tools that target such devices should implement functions using the fewest number of logic modules possible. Sometimes the fastest design is also the smallest in terms of number of modules used. In other cases, the synthesis software can introduce parallel logic to reduce the number of logic levels and improve performance.

One of the keys to good synthesis software, and one of the most difficult to achieve optimally is the efficient use of logic modules in synthesizing logic. Tools must be capable of searching a library of cells and select those that provide the highest number of gates per logic module. The ability to conduct such a search depends on the algorithms the synthesizer software employs.

In a discreet PLD, all the IOs connect to other devices via PCB traces. When the PLD logic is incorporated into an FPGA as part of a larger design, most of the IOs are connections internal to the device. Typically the PLD design is a state machine or counter whose outputs drive controls of data path logic such as multiplexors or registers which are also implemented in the FPGA.

Actel devices use multiplexors rather than three-states when a design requires multiple drivers for a line. There is a slight difference in migrating PLD designs to an FPGA depending on whether the three-state in the PLD was used or not. If the PLD three-state control is a pin and was not used in the original design, simply changing the pin's name in the pin list to NC will signal the Actel synthesis tool to ignore the unused PLD three-state functionality. The synthesized logic outputs can drive other logic in the FPGA. If the design requires that the synthesized logic outputs drive a three- state IO pad, simply assign a signal name to the three-state pin position in the pin list.

For PLDs with internal three-state control, such as the 22V10, the situation is analogous to a three-state control pin. The logic controlling the three-state pin on the PLD can be synthesized to drive a three-state IO pad on the FPGA, or to be an internal output to control a multiplexor select line.

Multiple PLD outputs can be used to drive a line if the internal three-state control lines are activated by mutually exclusive logic functions. An example of such an application may be seen in Figure 3.6.6 where an equation defines the three-state output enable logic. Like the other logic in the PLD, the three-state output enable control is entered in a sum-of-products format and implemented in the PLD AND-OR array.

When migrating a discreet logic design which includes PLDs to an FPGA all the logic

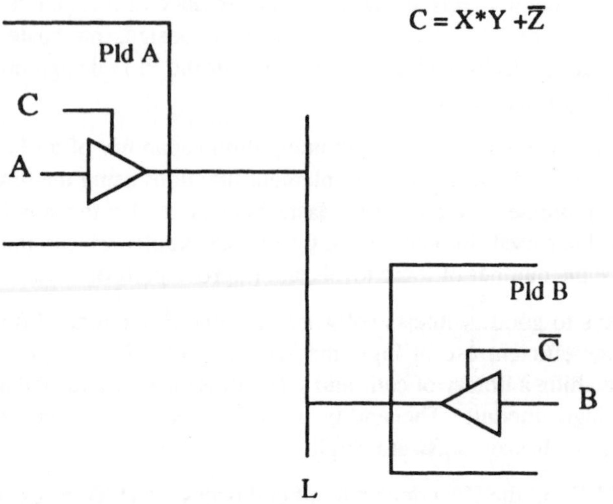

Figure 3.6.6. PLD Three State Output Enable Logic

is typically in a single device. Internal three-state functions may not be available, but even if available they tend to be slow.

An alternative to the internal three-state would be to use a multiplexor as shown in Figure 3.6.7 . A logic synthesizer can replace the three-states with the multiplexor and synthesize the output enable control logic to drive the multiplexor select line.

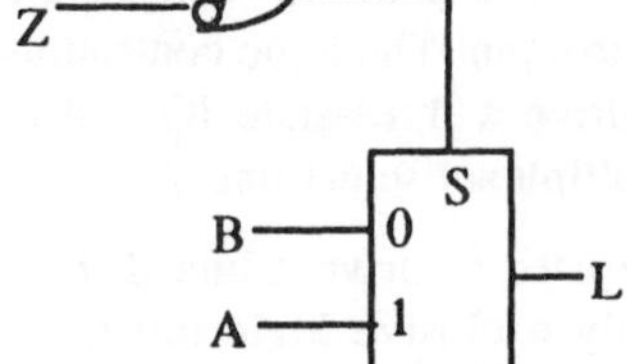

Figure 3.6.7. FPGA Multiplexor Alternative

Synthesis Design Flow

The design flow for the Actel Synthesis is shown in Figure 3.6.8 The designer may

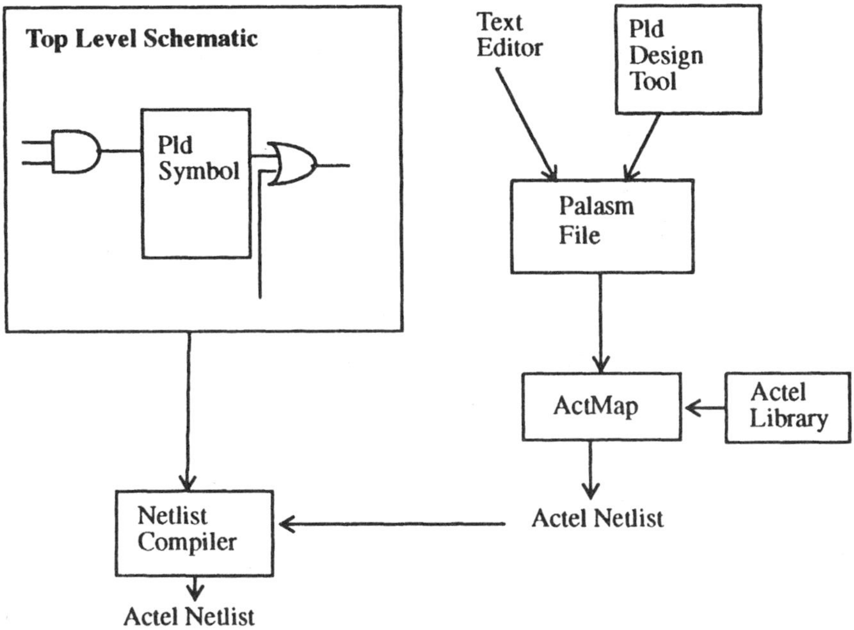

Figure 3.6.8. ACTMAP Design Flow

use any design entry tool that can output a PALASM-format file. Popular tools that can output the format include ABEL, CUPL, PLDesigner, and LOG/iC. Alternatively, a PALASM file may be written using a text editor.

After taking the input file, the synthesizer will generate a suboptimal netlist and proceed to optimize it. The user may influence the optimization process by using command line arguments to optimize for minimal capacity or maximum performance. The latter argument stipulates a performance goal that the tool tries to meet.

Since the actual delays for the complete design aren't known, the synthesizer assumes a typical routed delay value. If it cannot meet the criteria, it will report the performance it did achieve. The resulting netlist is in the Actel ADL format and resides in a separate directory.

The designer must create a symbol to be instantiated in the schematic that represents the synthesized netlist. The symbol IOs correspond to those defined in the PALASM file. It is marked with a property indicating that the netlist was synthesized and is to be found in a separate directory. When the top-level design is compiled, the schematic and the synthesized netlists will be integrated into a single file.

The ACTMAP tool also generates a netlist in the format of the users CAE tools which may be used to simulate and verify the synthesized logic either alone or interacting with other logic in the design. Schematic generators may be used to view the synthesis results.

Designing Counters with ACT Devices

Perhaps the most common digital logic function used is the synchronous binary counter. Regardless of the technology employed to implement counters, they are found in every type of application. The following describes some of the techniques for designing counters, or modifying a counters in the Act1 and Act2 soft macro libraries.

The functional requirements of the counter determine the approach to be taken in its design. Most applications minimally require synchronous counters with an asynchronous reset.

A simple binary counter with reset is shown in Figure 3.6.9 . The least significant bit uses an inverting flip-flop so it toggles without additional logic. Whenever counter output polarity is flexible, or parallel flip-flops are warranted by fan-out or performance considerations, it allows the design to take maximal advantage of macro bubbled IOs to reduce module count and levels of logic.

In the simple counter, as in most types of counters, the biggest design concern is propagating flip-flop outputs to higher bits. For Act1-based designs like the simple counter example, bits higher than eight require another level of combinatorial logic. Combinatorial levels required for the simple counter are: two for up to nine bits, three for ten to 33 bits.

A loadable counter appears in Figure 3.6.10 . The multiplexor flip-flip is used where one input provides the loaded data and the other the count data. The multiplexor select line is controlled by the counter load line.

The Act2 modules offer more logic integration than Act1 modules. In the sample design of Figure 3.6.11 you can see how to construct a six-bit loadable counter with only one level of combinatorial logic between flip-flops. (In the counter design described below a ten-bit counter with one level of combinatorial logic may be built using a multiplexed flip-flop with a gated select input.) The design has a 4:1 multiplexor driving the data input of each flip-flop. The multiplexor macro and the flip-flop are combinable into a single S-module by the ALS software providing good integration and performance.

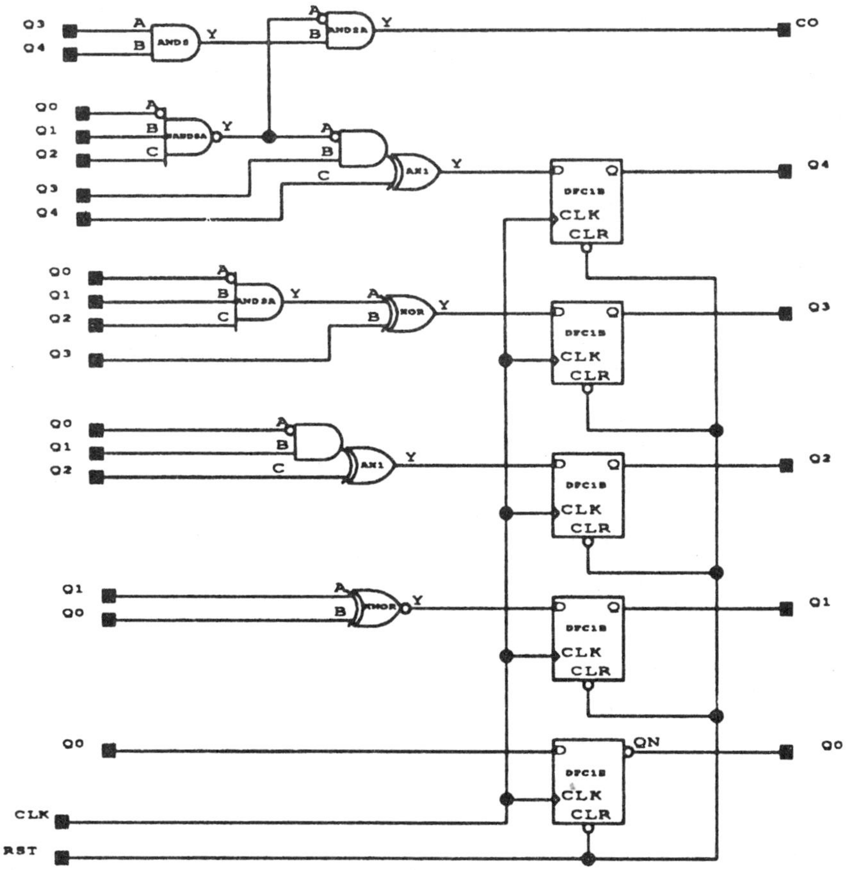

Figure 3.6.9. Act1 Five Bit Binary Counter

The select lines on the multiplexor are operated by the load control (S1) and by the count enable and carry from the lower order bits (S0). The multiplexor data inputs are used for data to be loaded, held, or incremented. The fourth multiplexor data input can be used as a second data input, or a synchronous set or clear.

Large counter design considerations should be evaluated from the perspective of the device-level design and how the counter is used in it. For example, if some counter outputs may be active low or if additional modules are used for redundant logic (e.g. some bits have both an active-high and an active-low output), then larger counters may be designed without additional logic levels using five-input gates. Such decisions should consider the implications for module count and fan-out as detailed below.

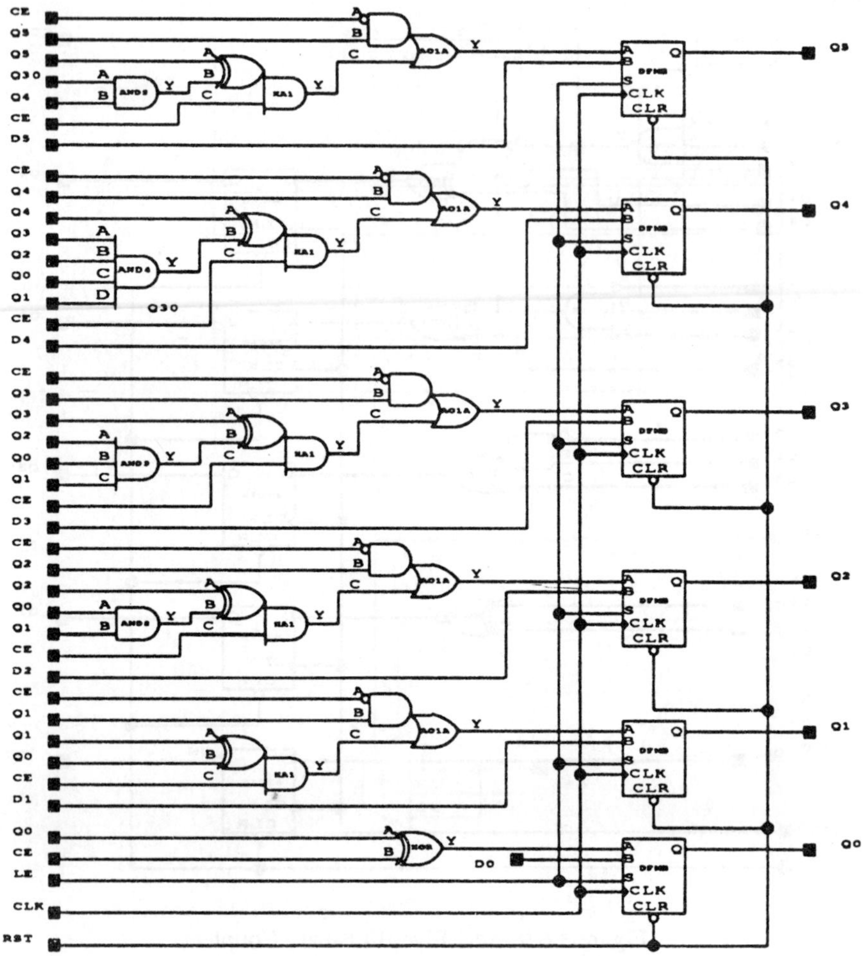

Figure 3.6.10. Act1 Loadable Counter

The sample design for a 16-bit synchronous loadable binary counter with a count enable will serve to illustrate some of the considerations designers should be aware of when designing large counters in Act2 FPGA designs. The functional description for the counter appears in Table 8

Table 8: Counter Function

Reset	Load	Counter Enable	Clock	Q
0	X	X	X	0
1	1	X	Rising	D
1	0	1	X	Q
1	0	0	Rising	Q+1

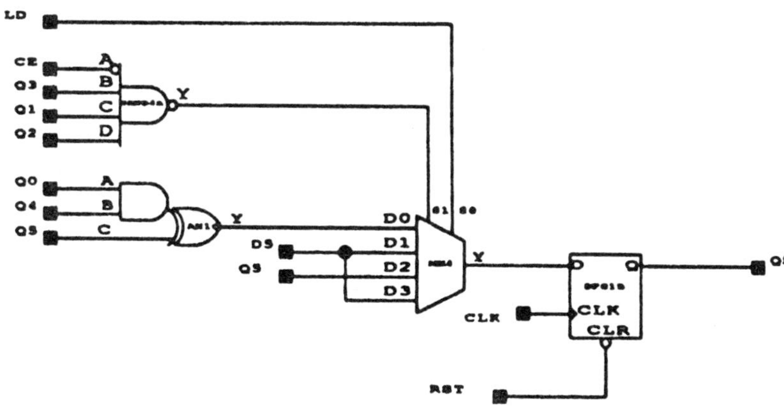

Figure 3.6.11. Act2 Six-Bit Loadable Counter Example

The counter design makes extensive use (bits 0 through 5) of a 4:1 multiplexor driving a flip-flop as depicted in Figure 3.6.12 . C-modules are used as AND-EXORs and for ANDs to qualify the count function. The AND function is also used to bring the count enable (*CE*) to the multiplexor via the select line in bits Q3 and above. Using both the select inputs as well as the data inputs to AND lower order bits allows for more paralleling resulting in fewer levels of logic.

For the bits Q6 through Q15, an S-module macro with both a 4:1 multiplexor and an OR gate driving one select line is used to allow for more parallel propagation of the lower bits. Figure 3.6.13 shows the implementation of the most significant bit of the

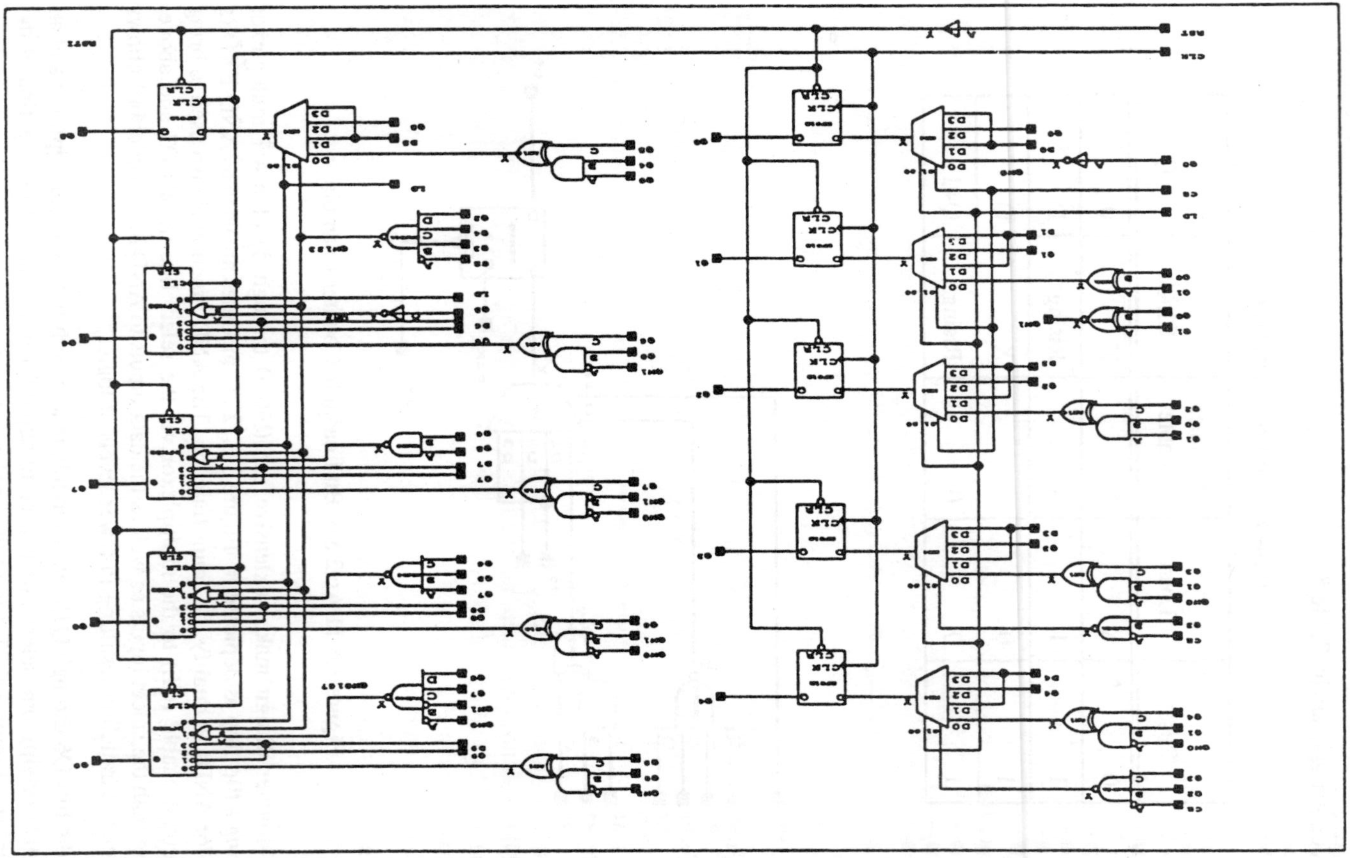

Figure 3.6.12. Counter Example Using Multiplexor And Flip-Flops

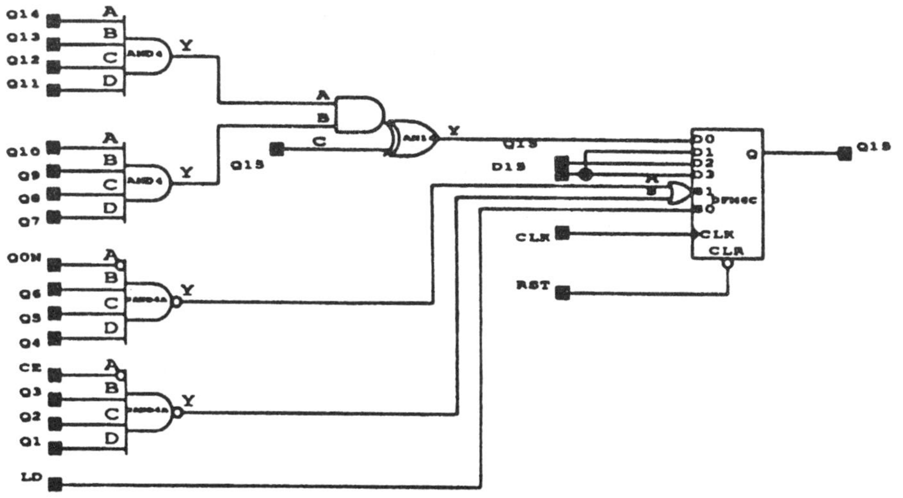

Figure 3.6.13. Counter Most Significant Bit

counter.

The S-module OR gate is used as a two-input NAND with active-low inputs which are, in turn, driven by NAND gates to propagate the lower bits and the count enable. The active-low output of the built-in two-input gate (OR used as a NAND) is adjusted for by shifting the position of the multiplexor data inputs.

Most four-input gates are implemented with a single C-module, but a four-input NAND with no bubbled inputs requires two modules. The limitation is avoided by using a NAND with a bubbled input. The count enable is active low, so it may be used to drive a bubbled input on a gate. Active-low counter bits are used to drive other bubbled gate inputs.

As may be seen in the schematic of Figure 3.6.12 , two bits (Q0 and Q6) use inverters. The Q0 inversion toggles the flip-flop. A toggle flip-flop could have been used instead of a D flip-flop, but it could not have been combined with the multiplexor into a single S-module. Moreover, the inverter output is available as a resource to share the fan-out load with the flip-flop and to allow the use of bubbled inputs on gates whenever it is desirable.

It could be argued that the use of the inverted output to drive gates causes the lower level bits to use two levels of combinatorial logic when it is not necessary. For a design of ten bits or less the point would be valid because no path requires more than one combinatorial level. In the example design, however, two levels are already required by the upper bits and the improvement in fan-out from the use of the inverter output at no additional cost in module count makes the practice worthwhile.

The paths in the design that are most likely to limit performance are those with the largest number of logic levels and the highest fan-out. In general, when fan-outs exceed nine on a critical path, using redundant logic is often clearly called for. For lower fan-outs the decision to use redundant logic is problematical and must be balanced by considering both the cost in additional logic modules as well as the fan-out to the outputs driving the redundant logic.

In the sample design, two redundant modules were added to illustrate the concept. One is an XNOR gate whose output is the inversion of Q1. The other is a four-input NAND gate which propagates flip-flop outputs. No fan-out in the design exceeds seven and the worst-case path is the redundant gate whose inputs are driven pins with fan-outs of seven, six, six, and five; and whose output fan-out is four. A total of 48 modules were used in the design.

Pre-scaled Counter Design

An important realization in designing high-performance counters is the fact that the least significant bits of the counter change the most frequently, and higher order bits change much less often. This fact can be used to optimize counter performance by making sure the least significant bits propagate up at the fastest rate. Higher order bits have a longer time to propagate through the logic. Lets consider the design of a loadable six-bit counter using the technique. It will be a down counter, suitable for timing or address generation.

The counter needs to have a least significant bit which can toggle at the highest possible rate. In the Act2 family the sequential logic module allows a 4-input multiplexer with gated select lines and a D-type flip-flop to be implemented in a single level of logic. The logic module can be used to construct a least significant counter bit with clear, load and count enable as shown in the upper portion of Figure 3.6.14 .

Data can be loaded into the register when the load enable (LD) signal is high. The count enable signal (CNT) is used to toggle Q0 when it is active and hold Q0 when inactive. The circuit implements a least significant bit which needs only a single level of logic to operate.

Figure 3.6.14 shows the implementation of Q1 using the Act2 module. It is similar to the LSB except Q1 is inverted to toggle the flip-flop. Since Q1 can only toggle when Q0 is a zero (in a down counter), we will have an extra clock cycle for the inversion to propagate for Q1's next state. That allows the slower /Q1 time to settle, insuring fully synchronous operation. The basis of the design technique will be to have slower signals gated off until they have had sufficient time to settle and they are needed to compute the toggle of the associated counter bit.

Added to Q0 and Q1 are three registers to source CNT, /CNT and LD as shown in the lower part of Figure 3.6.14 .The complete schematic illustrated Figure 3.6.14 show a

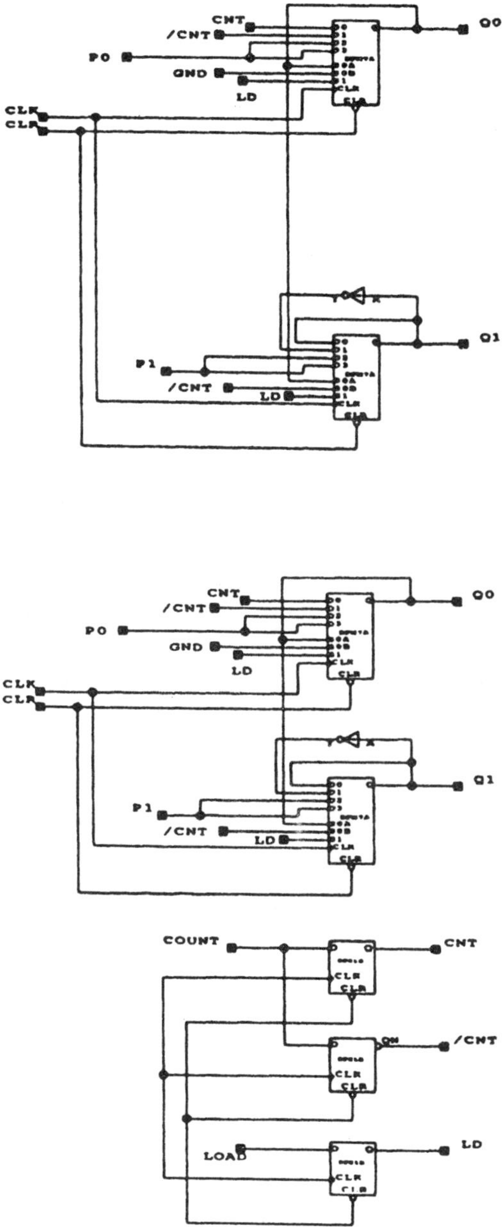

Figure 3.6.14. Act2 Pre-scaled Counter

2-bit pre-scaler (later referred to as CNT2P) for fast counters.

A four-bit macro for the MSBs of the six-bit counter is shown in Figure 3.6.15 . It

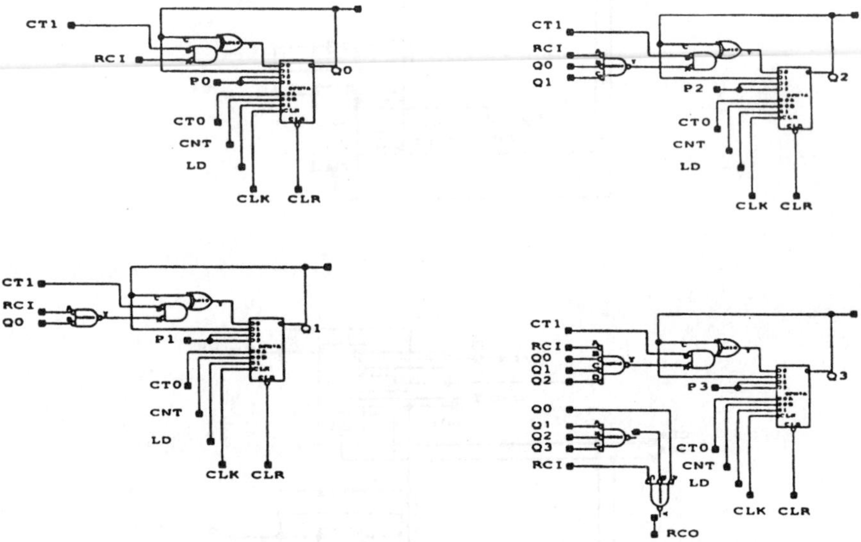

Figure 3.6.15. Four Bit Counter Macro

uses the LSB from the CNT2P macro (connected to CT0) and CNT to enable the multiplexer inputs used for toggling. The next LSB (Q1) from CNT2P is connected to CT1 along with the counter bits in the macro (Q0-Q2) to gate the input to the XOR determining the next count. A ripple carry input (RCI) is also used to allow results from previous stages to participate. This technique allows the MSBs four clock cycles (the time for the LSBs to transition from 0 through 3) to develop the input to the associated bits in the counter. The complete six-bit counter is shown in Figure 3.6.16

By replicating the pre-scaler, fan-out limited counters can be constructed. As an example, an 18-bit counter implemented with the above macros is shown in Figure 3.6.17 . The limiting frequency for the counter is determined similarly to the six-bit

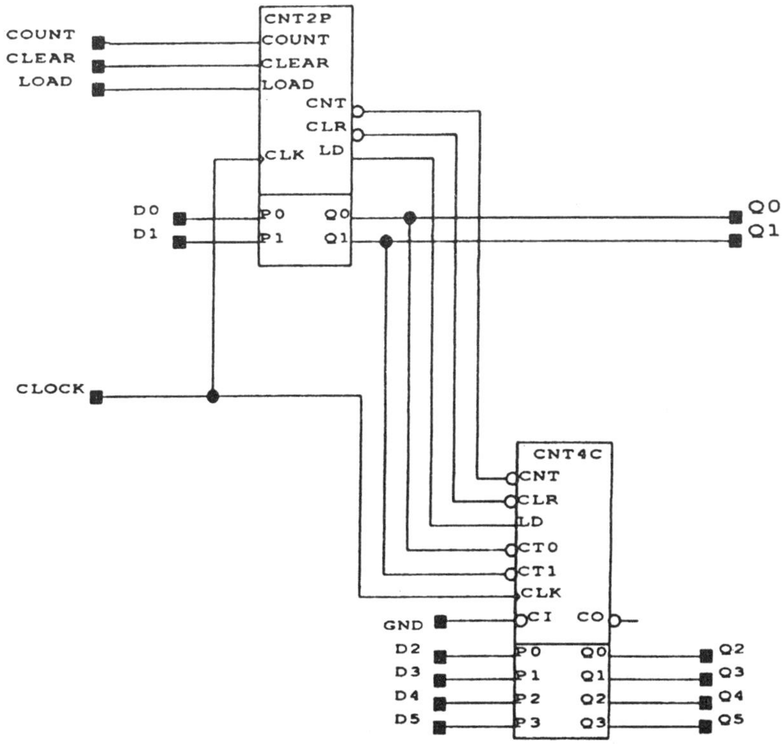

Figure 3.6.16. Six Bit Counter

example.

Designing Adders and Accumulators with the ACT Architecture

Many designers implement adders using carry-propagation techniques. The multiplexed-based Act1 and Act2 combinatorial module (C-module) allows for the more efficient carry-select design. This method partitions the add function into blocks that perform two additions simultaneously on a number of bits of the two operands.

The two additions are performed simultaneously except that one assumes a carry-in and one has no carry-in. The two sums are input to a 2:1 multiplexors, one for each bit pair. The carry line, from the low bits to the high bits, is used to select the appropriate sum for each block.

The ACT architecture lends itself well to implementing adders of various sizes using

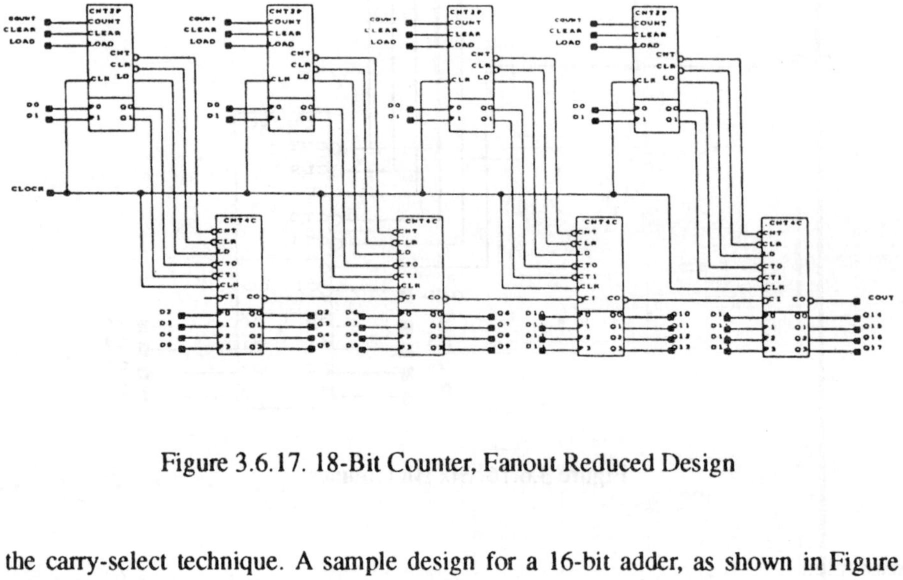

Figure 3.6.17. 18-Bit Counter, Fanout Reduced Design

the carry-select technique. A sample design for a 16-bit adder, as shown in Figure 3.6.18 , will be used to illustrate adder design.

The method for obtaining optimal performance from a carry select adder is to design it such that the number of levels of logic required for the carry chain equals the number for the largest sum block as closely as possible. When they are the same number of levels, the sum bits arrive at the data pins and the carry arrives at the select pins of the output multiplexing stage simultaneously.

The way to balance the levels of logic modules for the sum blocks with the carry is to partition the sum blocks. This partition is based on the logic levels required for the sums and the levels for the carry between sums. The size of the partitions varies with width of the data. The Act2 library contains some powerful hard macros that are used to shorten the levels of logic required for generating sums and carries. The description of the sample design will illustrate the use of the macros.

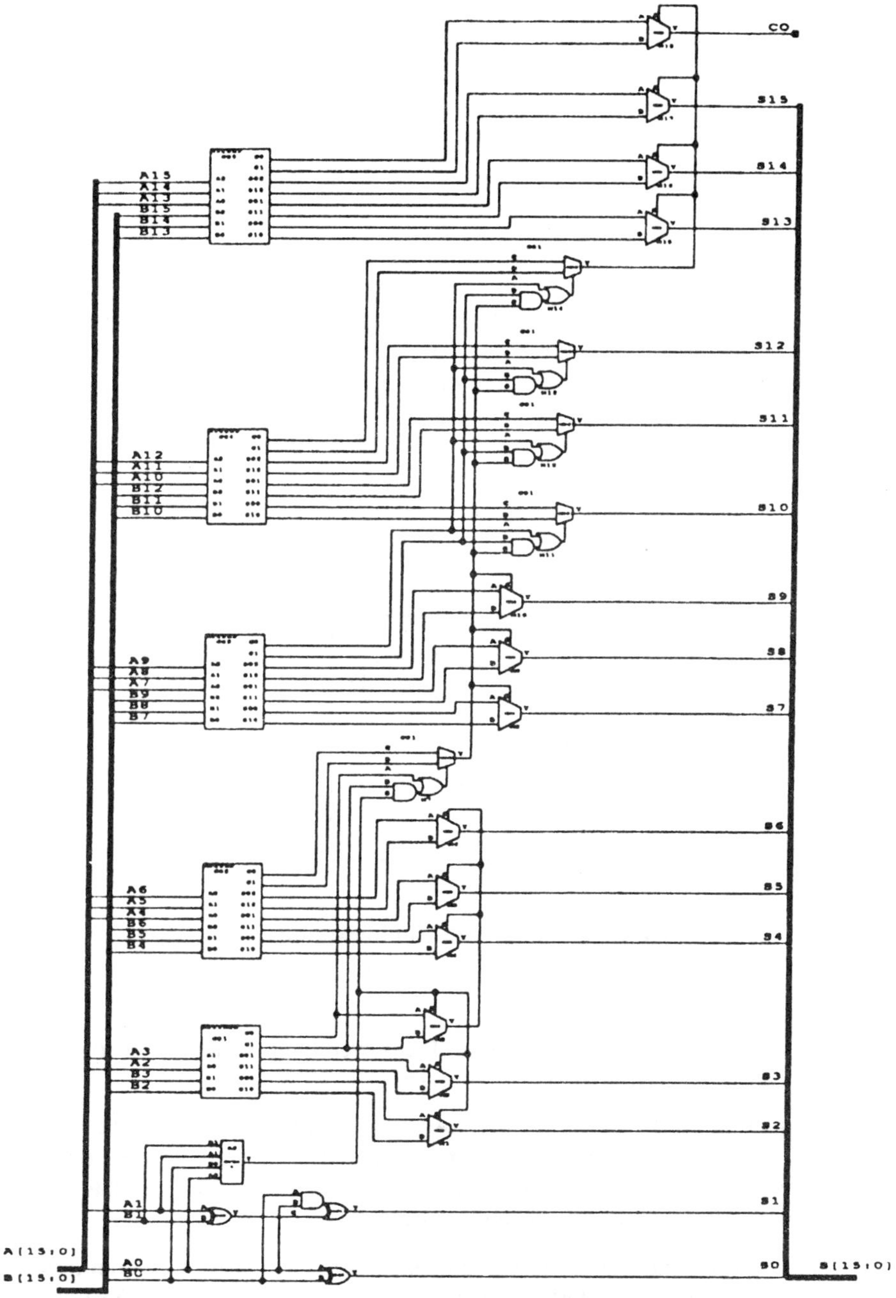

Figure 3.6.18. Act2 Carry-Select Adder Example

For the 16-bit adder, the optimal organization is to perform two two-bit additions on the least four significant bits with the remaining higher order bits broken into four sections of three bits each. In the top-level schematic the addition logic of the two least significant bits is visible. The other additions are performed in lower levels of the design hierarchy described in the next section.

The Act2 library includes two two-level carry hard macros. One macro generates a carry for the two bit pairs assuming the carry-in is true and the other assumes it is false. The latter macro may be seen at the bottom of Figure 3.6.18 making the carry for the least two bits.

The carry macro output drives the select line for the 2:1 multiplexors for sum bits two and three. It also drives the select line on the cascade multiplexor. The cascade multiplexor is a special Act2 hard macro that can propagate two levels of carry. The macro is depicted in Figure 3.6.19 and has five inputs. The top multiplexor inputs

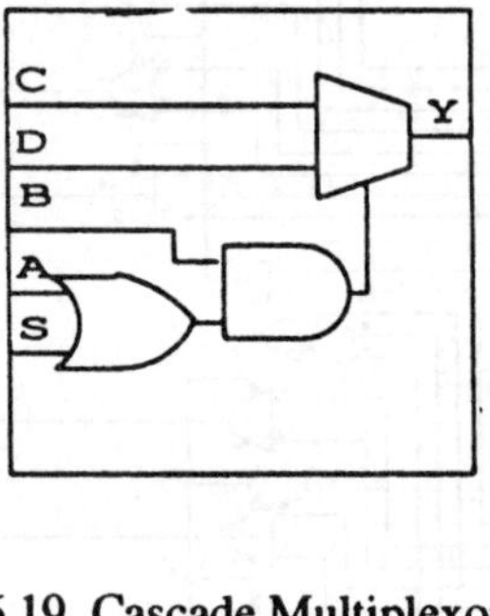

Figure 3.6.19. Cascade Multiplexor Macro

select the most significant sum or carry. The lower three inputs drive logic that implements a simplified form of a 2:1 multiplexor.

A fully implemented 2:1 cascade multiplexor does not map into the Act2 module efficiently, but the full functionality is not required in a carry select adder. A simplified version of the cascade multiplexor that maps into a C-module or that can be combined with a flip-flop in an S-module is available.

The simple version has logic driving the select for the upper level multiplexor consisting of only a two-input OR driving one input of a two-input AND. The two OR gate inputs are driven by the carry output from the next lower sum block assuming no carry-in and the carry in from the rest of the lower bits of the adder. The remaining AND input is the carry from the sum block which assumes a carry-in.

The logic is correct for a carry select adder because if the assume-no-carry-in input is true (meaning that a carry was generated within that sum block), then the assume-carry-in is always true (since it equals the false plus one) which completes the AND function.

If the carry from the lower bits is true (meaning a carry is propagated to the sum block), then we complete the AND if the assume-carry is true.

The schematic for the three-bit adder block appears in Figure 3.6.20 The adder

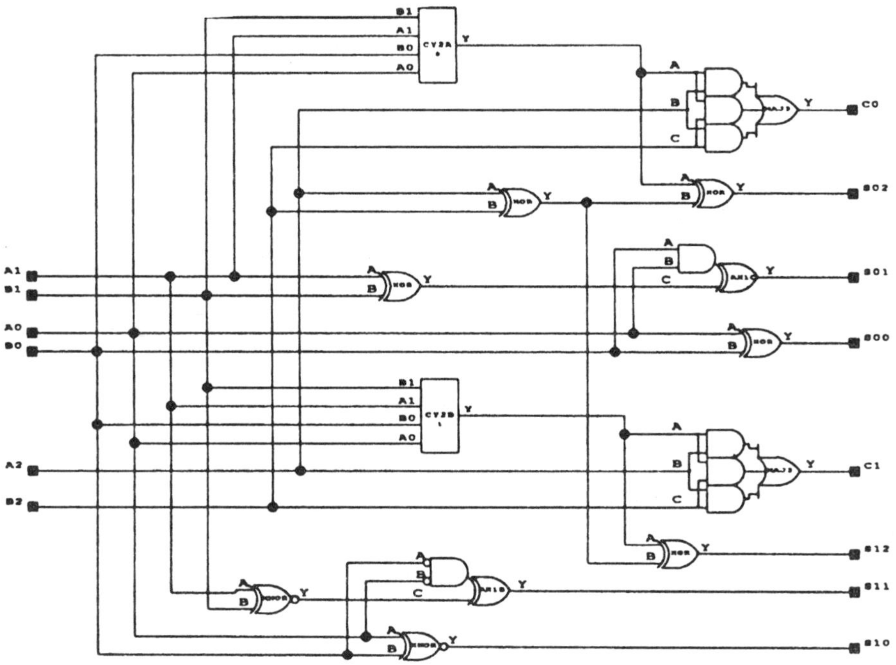

Figure 3.6.20. Three Bit Adder Block

requires thirteen logic modules to generate the three sum and carry pairs. All the output paths are two levels of logic or less. The two carries for the three bits come from two-level carry hard macros driving a three-bit majority macro. All the sums are generated from exclusive OR or NOR gates.

The optimal design for a carry select adder depends on the number bits to be added. As mentioned previously, the number of bits in a block is a function of the overall adder size. For Act2 adders of two and three bits, the design shown in Figure 3.6.20 is attractive because the two-level carry macro allows for delays to be two levels or less.

An example for another design for an adder block is shown in Figure 3.6.21 . This

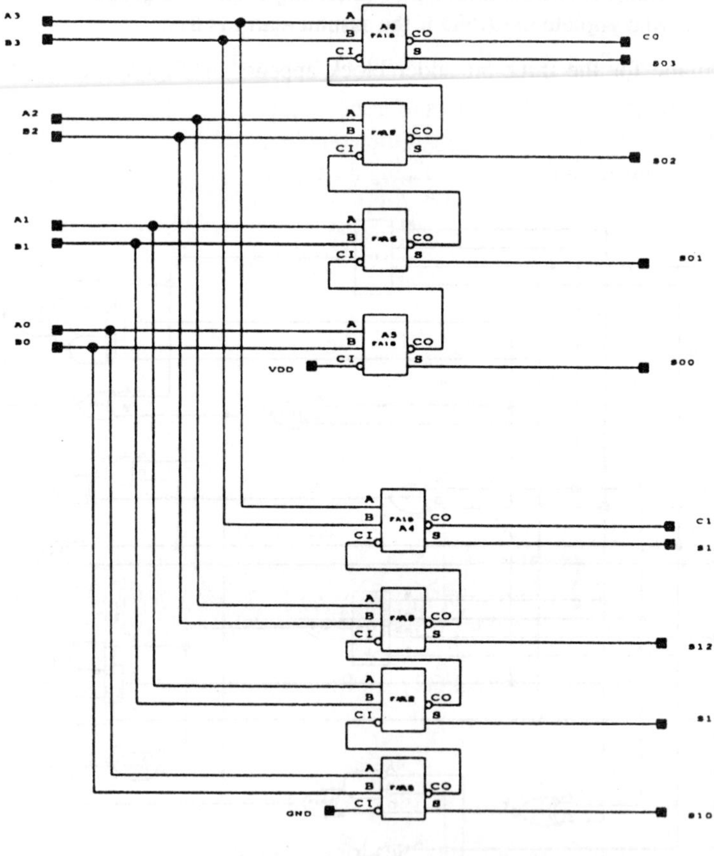

Figure 3.6.21. Another Adder Example

design uses ACT library single bit two-module hard macros. The Actel library contains several such two-module adder macros with both a sum and a carry output. The carry output goes through one module delay and the sum output two module delays. The carry propagates through the chain in four levels in the example so that all the sum bits are stable when the carry ripple is complete.

A three bit adder block has one level of delay more than that of Figure 3.6.20 , but requires one less module. For Act1 adder blocks and Act2 blocks greater than three bits, the carry chain design is most efficient.

Accumulator

All the sum and carry outputs of the adder macro are combinable into a single Act2 sequential module (S-module). This combinability feature means that if the data inputs of a 16-bit register are schematically connected to an adder's outputs, the ALS software will automatically put the adder output macros (2:1 multiplexors or cascade multiplexors) into their respective flip-flop in the register.

The registered-output adder will suffer no degradation in performance from the combining because the delay through the combinatorial part of the S-module is less than that of an uncombined macro. Tieing the register output back into the inputs will make the circuit into an accumulator. A sample design for an accumulator made from an adder and a register may be seen in Figure 3.6.22 .

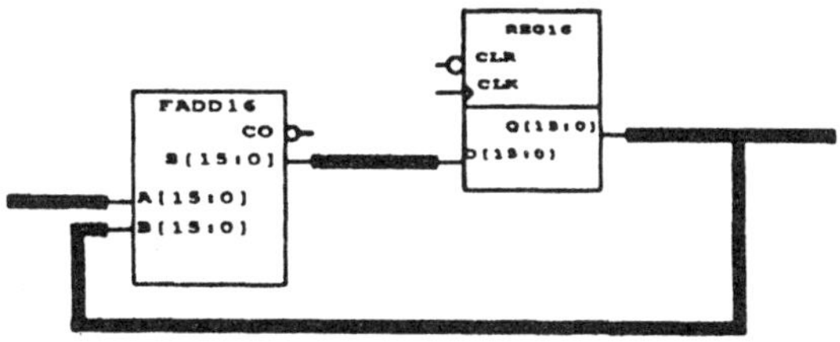

Figure 3.6.22. 16 Bit Accumulator

Design Results

The sample design uses 82 Act2 modules for either the adder or the accumulator. The slowest path in a function is usually the one with the most levels of logic. In this case it is the carry chain which has four levels of logic. As mentioned previously, all other paths have fan-outs of three or less.

The modules in the chain have fan-outs of three, four, seven, and four. Criticality may be used to optimize the path performance. Criticality works best when fan-out is low. When the fan-out of a speed-sensitive net exceeds seven, performance can usually be most improved by adding redundant logic. For fan-outs of less than seven, adding redundant modules may not bring any improvement. Using redundant logic for fan-outs of seven should be considered on an individual basis. Adding a redundant module to the carry path would change its fan-outs to three, five, three, and four. The expense of one module may be justified by the performance improvement from lowering the fan-out.

It is also possible to improve performance by pipelining an adder. Since all of the combinatorial functions used in the adder are combinable (if the function's output drives a flip-flop ALS will put both in a single S-module), designers may pipe the adder at the points that provide the best performance at no cost in additional modules.

The carry select architecture is extensible to adders of any size. Adders of eight to fifteen bits may be designed using the technique in three levels of logic. Adders from 16 to 24 bits can be done in four levels. When adapting the adder design to other operand sizes remember to re-partition the sum block sizes to match the logic levels of sums and carries.

State Machine Design

The traditional methodology for designing state machines has been to draw a state diagram, map the states into the minimum number of register bits needed to encode them and determine the minimal next state function. The method results in a minimum number of registers, but often requires wide gating to determine the next state bit because all the register outputs are feedback to each transition term.

PLDs are register lean but can do wide gating, so their architecture fits into the above methodology easily. FPGAs however offer designers a different set of resources. The style of implementation of the state machine output coding should consider how best to take advantage of the resources.

On Act1 devices flip-flops use two modules and are relatively expensive compared to gates. For these devices, encoded outputs are often most efficient. On Act2 devices, registers are abundant and gating is optimized for more narrow functions. A state machine designed to take advantage of the large number of registers available on an Act2 FPGA may be more efficient than the PLD implementation. A sample state machine illustrated in Figure 3.6.23 will be used to contrast the encoded state method with the Bit-Per-State (BPS) technique which uses a single register for each state. We'll use the encoded method to implement the design using the Act1 library and repeat the process using BPS.

The example contains six states, seven inputs, and five outputs. The traditional approach would encode the states into three state bits by making state assignments. The schematic in Figure 3.6.24 shows one possible implementation. The logic required to implement the state machine consumes 29 Act1 modules. The longest path is through four levels of delay.

With register rich ACT devices BPS becomes more attractive. State assignment is

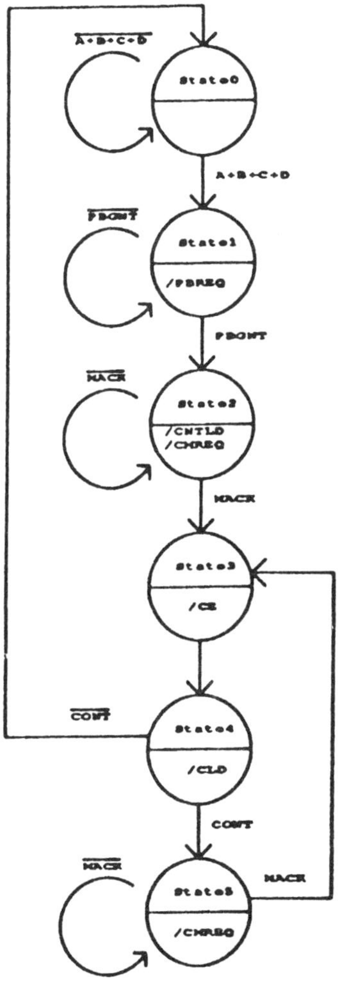

Figure 3.6.23. Example State Diagram from 4-Channel DMA Controller

trivial if a register is used for each state bit and the design will directly implement the transition terms. The schematic of Figure 3.6.25 exemplifies the results of the BPS method. It uses 19 Act1 modules and only requires two levels of delay.

Note that the state bits are inverted to make the states active low and the single

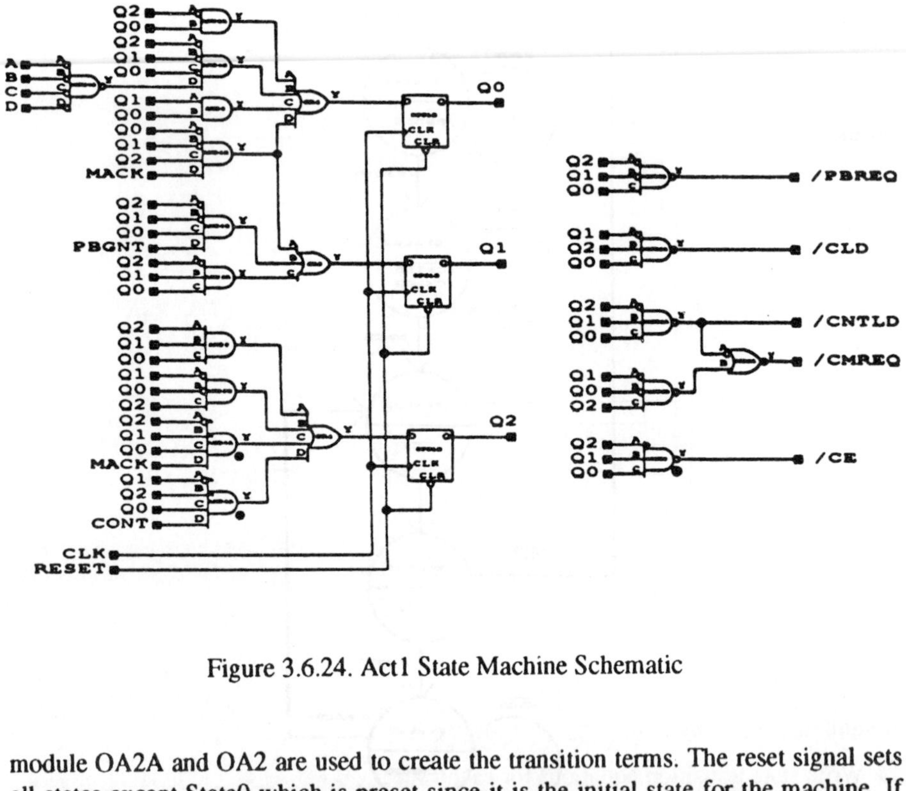

Figure 3.6.24. Act1 State Machine Schematic

module OA2A and OA2 are used to create the transition terms. The reset signal sets all states except State0 which is preset since it is the initial state for the machine. If active high state bits are required they can be used by clearing them on initialization and detecting the active high state for transition purposes.

Larger state machines may also be implemented using this technique by distributing control to several smaller machines and using a single master machine to coordinate activities. This usually results in a higher performance design since control signals will be located near the logic they affect, minimizing routing delays. In addition, typically it is easier to design and debug since each machine can be more easily understood and interactions between operations are minimized.

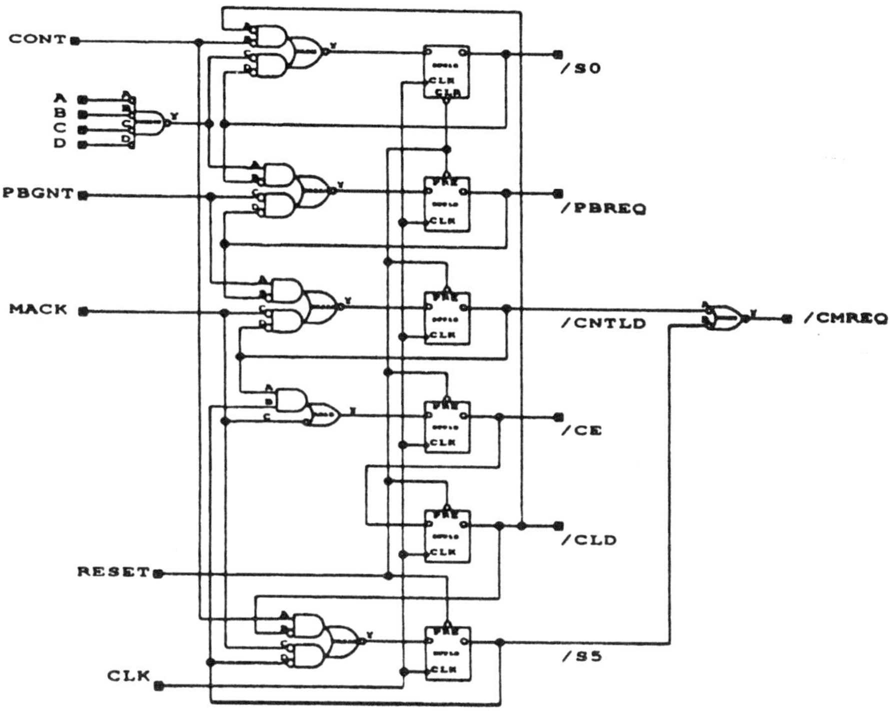

Figure 3.6.25. Act2 State Machine Implementation

A summary of the methodology is as follows:

1. Write state transition equations for each state.
2. Assign each state to a separate register. Where possible state bits should be made active low to make it easier to construct transition terms in a single ACT 1 module.
3. Take output signals directly from state outputs where possible. If the output signal must be active high, the state bit associated with it can be made to correspond by inverting its driving logic, and inverting its inputs to any other transition term.

4. All active high state bits are reset on machine initialization and active low states are preset on initialization. The initial state is must be activated on initialization so it should be cleared if active low or preset if active high.

Using FPGAs for Digital PLLs

In addition to purely digital applications, many designs use FPGAs for Digital Signal Processing (DSPs). We'll examine one such application, digital PLLs, to show two ways of implementing PLL designs using FPGAs.

Pulse Steal PLL

In telecommunications applications it is often desirable to generate a digital signal which is locked to an incoming signal and is some multiple of its frequency. A drawing of a pulse steal PLL which is a simple way generate such a signal may be seen in Figure 3.6.26 . Note that the design contains an ordinary oscillator, but no VCO. Except for the crystal, the entire design will operate in an FPGA.

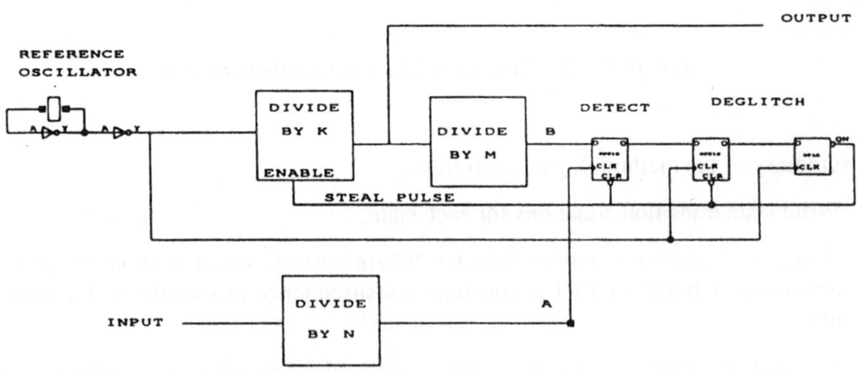

Figure 3.6.26. Pulse Steal PLL

Note the frequency relationship that holds at points A and B in the figure where:

$$OSC/(K*M) = INPUT/N = COMPARISON\ FREQUENCY \quad \text{(EQ 1)}$$

This technique is based on selecting a reference oscillator frequency that is slightly higher than OSC. This frequency (OSC+) should be chosen so that:

$$1/COMPARISON\ FREQ. - (K*M)/(OSC+) = .5 * (1/OSC) \quad \text{(EQ 2)}$$

The right side of EQN. 2 equals one half the period of the reference oscillator.

The reference oscillator frequency delta will cause point B (the detector flip-flop D input) to begin to precede point A (the detector flip-flop clock input) by half a period. When the edge of the D input is sufficient the detector will clock true and begin a pulse train through the two de-glitching flip-flops. The output of the second of these clears all three flip-flops and steals a pulse by disabling the divide by K output. Stealing the pulse puts point B behind A until the reference oscillator delta can move it ahead by one period repeating the cycle. Points A and B are always within one half a cycle of each other.

The circuit allows the frequency of the output signal to be selected simply be adjusting the values of the dividers K and M. The lock range of the loop is given by the following:

$$Lock\ Range = +- (OSC+/osc)/INPUT \quad \text{(EQ 3)}$$

Jitter Bounded PLL

Another technique [Walters] for generating a wide variety of synchronized clock frequencies with low jitter employs an accumulator Digital Controlled Oscillator (DCO) and phase and frequency comparators. The system, shown in Figure 3.6.27 , can lock to any division of a reference frequency (F_{ref}), as selected by the data loaded into frequency divider counters.

The Successive Approximation Register (SAR) and its controller serves as a low-pass filter supplying the DCO with frequency and phase correction data. Among the three inputs to the SAR controller is the F_{ref} divided by a factor Q to form F_q.

The other two inputs come from the phase and frequency (zero) comparators. The frequency comparator output is the DCO frequency divided by P to form F_p. When the system is in lock the following equation is true:

$$F_{dco} = (P/Q) * F_{ref}. \quad \text{(EQ 4)}$$

The heart of the system is the accumulator DCO which determines the ability to lock to a frequency and the amount of jitter allowed. The DCO consists of a four-bit accumulator whose input is fed by the SAR. The DCO input value is determined from

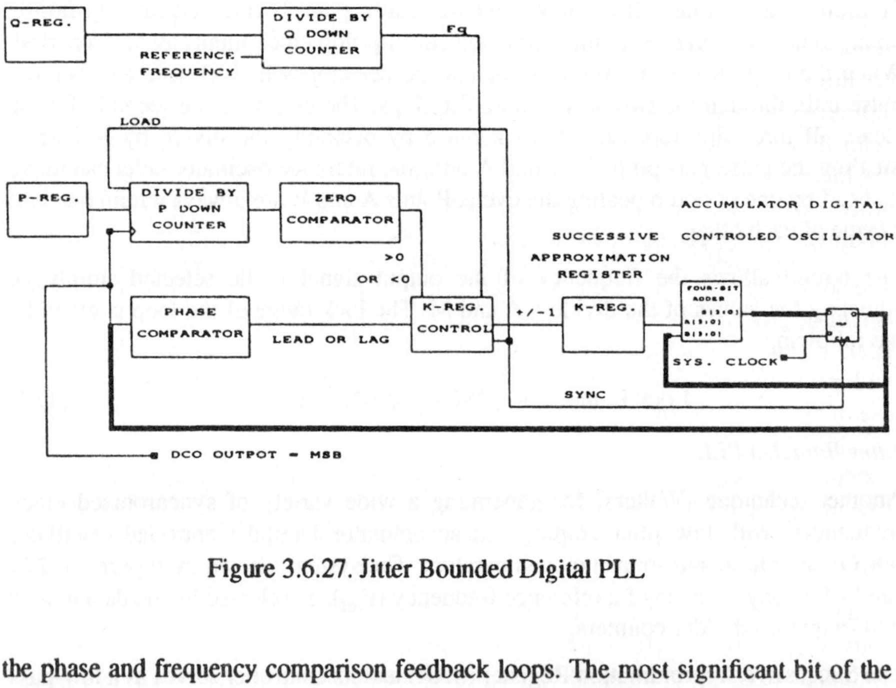

Figure 3.6.27. Jitter Bounded Digital PLL

the phase and frequency comparison feedback loops. The most significant bit of the accumulator output is the DCO output signal. It is generated by successively adding the SAR value to itself at the high- frequency system clock rate. The frequency comparator uses the value of P to divide the DCO frequency. If the frequency is out of lock during a period of $F_{ref.}$, the comparator asserts greater-than-zero or less-than-zero to the SAR controller to modify the value of the register. If the P counter output is zero, the DCO has the correct frequency.

The DCO latch acts as a phase register indicating the phase of the DCO with respect to F_{ref}. The DCO phase is calculated by the N most significant accumulator output bits. When the DCO is out of phase the jitter, or phase difference, is detected by the phase comparator and accumulates with time until it equals one period. The feedback loops then cause the SAR register controller to load a correcting value into the register or to clear the accumulator with a synchronizing pulse.

The Jitter-Bounded DPLL may be implemented entirely on an ACT FPGA. The resource requirements vary with the relationships of the system input and output frequencies, but for any F ref, system clock, and desired output frequency the design is easily accommodated on an ACT FPGA.

Customer Design Examples

Example 1: Functional Prototyping

A multi-processor file server for a PC network is under development. The system employs multiple 80486 microprocessors, banks of DRAM, and multiple 64- and 32-bit busses. Several 1280s FPGAs are used to prototype the system which will ultimately be implemented with gate arrays. Since the logic was composed of clearly defined sub-blocks, principally state machines and simple data paths, partitioning of the logic among the FPGAs was not difficult. For this complex system, the design process is incremental and somewhat experimental. The logic in each FPGA may be modified or extended several times, so a fast design flow is necessary.

The logic for each FPGA is synthesized with the Synopsys hardware description language (HDL) compiler using an Actel macro library, and then checked with an HDL simulator. Each design is automatically placed and routed, and re-simulated with post-layout delays. Although the system will ultimately operate at a high speed, use of synchronous logic design allows the prototype to operate at a speed within the capability of the FPGAs. Software is developed and tested on the prototype concurrently with the completion of the logic design and conversion to conventional gate arrays.

Example 2: At-Speed Prototyping and Pilot Production

A digital video signal processing system in high volume production for the consumer market. Four 1020 FPGAs were used to integrate registers, multipliers, adders and digital phase- locked-loops. The pipelined data paths are 12 to 16 bits wide and operate at a 16MHz clock rate. For production the four FPGAs were replaced by a single 8000 gate masked gate array. It is estimated that the use of FPGAs for prototyping and pilot production saved two to three months in product development time.

As many as half of masked gate array designs never reach high-volume production. The reasons include short product lifetimes, required new features or other changes, or simple lack of demand for the product. In such cases use of FPGAs for production saves the NRE costs associated with developing a mask programmed gate array.

For very large volumes, the design may be transferred to a conventional gate array. Due to the small granularity of the module, an FPGA design can be efficiently converted to a gate array. Test vectors can often be generated automatically. Several vendors support this conversion path.

3.7 Acknowledgments

This chapter includes information compiled by many people. In particular Jonathan Greene and Esmat Hamdy contributed portions of this chapter. Sam Beal, Warren Miller and Jeff Schlageter both contributed material and made suggestions about the presentation.

3.8 References

[Ahrens] M. Ahrens, et. al., *An FPGA Architecture Optimized for High Densities and Reduced Routing Delay,* Proceedings Custom Integrated Circuits Conference, July1990.

[Allen] D. Allen, R. Goldenberg, *Design Aids and Test Results for Laser-Programmable Logic Arrays,* Proceedings. International Conference. on Computer Design, 1990, pp. 386-390.

[Birkner] J. Birkner et al., *A Very High-Speed FIeld Programmable Gate Array Using Metal-to-Metal Antifuse Programming Elements,* Proceedings Custom Integrated Circuits Conference, May 1991.

[Brayton] R. Brayton, R. Rudell, A. Sangiovanni-Vincentelli, A. Wang. *MIS: A Multiple-Level Logic Optimization System.* IEEE Transactions on CAD, Nov. 1987.

[Carter] W. Carter et al, *A User Programmable Reconfigurable Gate Array,* in Proceedings of Custom Integrated Circuits Conference, pp. 233-235, 1986.

[Chen] J. Chen et al, *A Modular 0.8 um Technology for High Performance Dieletric Antifuse Field Programmable Gate Arrays,* International Symposium on VLSI Technology, Systems, and Applications, May 1993, pp. 160-164.

[Chiang1] S. Chiang, R. Wang, J. Chen, K. Hayes, J. McCollum, E. Hamdy, C. Hu. *Oxide-Nitride-Oxide Antifuse Reliability,* International Reliability Physics Symposium., March 1990, pp.186-192.

[Chiang2] S. Chiang, K. Hayes. *Act 1010/1020 Reliability Report.* Actel Corporation, Sunnyvale, CA, April 1990.

[ElAyat] K. El Ayat, et. al. *A CMOS Electrically Configurable Gate Array.* IEEE J. Solid-State Circuits, Vol. 24, No. 3, June, 1989, pp. 752-762.

[ElGamal1] A. El Gamal, J. Greene, J. Reyneri, E. Rogoyski, K. El-Ayat, and A. Mohsen. *An Architecture for Electrically Configurable Gate Arrays.* IEEE J. Solid-State Circuits, Vol. 24, No. 2, April, 1989, pp. 394-398.

[ElGamal2] A. El Gamal, J. Greene, V. Roychowdhury. *Segmented Channel Routing is Nearly as Efficient as Channel Routing (and Just as Hard).* Proceedings. VLSI Conference, Santa Cruz, CA, March 1991.

[ElGamal3] A. El Gamal. *Two Dimensional Stochastic Model for Interconnections in Master Slice Integrated Circuits.* IEEE Trans. on Circuits and Systems, CAS-28, 127-138, Feb. 1981.

[Gerzberg] L. Gerzberg. U.S. Patent 4,590,589, 1986.

[Graham] H. Graham, D. Seltz. *Electronically Programmable Gate Array Having Programmable Interconnect Lines,* U.S. Patent 4,786,904, Nov. 22, 1988.

[Greene] J. Greene, V. Roychowdhury, S. Kaptanoglu, A. El Gamal, *Segmented Channel Routing,* Proceedings. Design Automation Conference., Orlando, Florida, Association. for Computing Machinery, June 1990.

[Hamdy] E. Hamdy, J. McCollum, S. Chen, S. Chiang, S. Eltoukhy, J. Chang, T. Speers,. Mohsen. *Dielectric Based Antifuses for Logic and Memory ICs,* IEDM Tech. Digest, pp. 786-789, 1988.

[Hashimoto] A. Hashimoto, J. Stevens. *Wire Routing by Optimizing Channel Assignment within Large Apertures.* Proceedings. 8th IEEE Design Automation Workshop, 1971.

[Holmberg], et al. U.S. Patents 4,499,557, 1985 and 4,599,705, 1986.

[Hsieh] H. Hsieh, et. al., *Third-Generation Architecture Boosts Speed and Density of Field-Programmable Gate Arrays,* Proceedings. 1990 Custom Integrated Circuits Conference., May 1990, pp. 31.2.1-32.2.7.

[Karplus] K. Karplus, *AMAP: a Technology Mapper for Selector-based Field-Programmable Gate Arrays,* Proceedings of the 28th Design Automation Conference, June 1991, pp. 244-247.

[Lim] Lim et al. U.S. Patent 4,569,121, 1986. Stacy et al. U.S. Patent 4,569,120. 1986.

[Lorenzetti] M. Lorenzetti, D. Baeder. *Routing.* Chapter 5 in Physical Design Automation of VLSI Systems, B. Preas and M. Lorenzetti, eds., Benjamin Cummings, 1988.

[LSI] *HCMOS Gate Array Databook and Design Manual,* LSI Logic Corporation, October 1986.

[Mailhot] F. Mailhot, G. De Micheli. *Technology Mapping Using Boolean Matching and Don't Care Sets.* preprint, 1990.

[Marr] C. Marr, *Logic Array Beats Development Time Blues,* Electronic System Design Magazine., Nov. 1989, pp. 38-42.

[Murgai] R. Murgai, Y. Nishizaki, N. Shenoy, R. Brayton, A. Sangiovanni-

Vincentelli. *Logic Synthesis for Programmable Gate Arrays.* Proceedings. 27th ACM/IEEE Design Automation Conference., 1990.

[Preas] B. Preas, P. Karger. *Placement, Assignment and Floorplanning.* Chapter 4 in Physical Design Automation of VLSI Systems, B. Preas and M. Lorenzetti, eds., Benjamin Cummings, 198.

[Roesner] B. Roesner. U.S. Patents 4,424,579 and 4,442,507, 1984.

[Rose] J. Rose, R. Francis, D. Lewis, P. Chow. *Architecture of Programmable Gate Arrays: The Effect of Logic Block Functionality on Area Efficiency.* IEEE Journal of Solid State Circuits, Vol. 25, No. 5, pp. 1217-1225, October 1990.

[Rudell] R. Rudell, R. Segal. *Logic Synthesis Can Help in Exploring Design Choices,* 1989 Semi-custom Design Guide, CMP Publications, Manhasset, NY.

[Schlageter] J. Schlageter et all, *An Advanced Sub-Micron Architecture for IO Intensive Applications,* Proceeding of the 1993 Compcon conference, 1993, pp. 362-366.

[Singh] S. Singh, J. Rose, D. Lewis, K. Chung, P. Chow, *Optimization of Field-Programmable Gate Array Logic Block Architecture for Speed,* Proceedings of the 1991 CICC Conference, 1991, pp. 6.1.1-6.1.6.

[Smith] J. F. Smith, et. al., *Laser-Induced Personalization and Alterations of LSI and VLSI Circuits,* Proceedings of. 1st International Laser Processing Conference., Anaheim, Calif., Laser Institute of America, Nov. 16, 1981.

[Stopper] H. Stopper, et al. U.S. Patent 4,847,732, 1989.

[Walters] Walters, S., Troudet, T.; *Digital Phase-Locked Loop with Jitter Bounded.,* IEEE Transactions on Circuits and Systems, VOL. 36, NO. 7, July 1989.

[Whitney] T. Whitney and J. Schlageter, *A New High Performance Field Programmable Gate Array Family,* Proceedings of 1993 International Conference on Computer Design, October 1993.

[Whitten] R. Whitten, R. Bechtel, M. Thomas, H.T. Chua, A. Chan, J. Birkner, European Patent Application No. 90309731.9, May 9, 1990.

[Wong] S. Wong, H. So, J. Ou, J Costello, *A 5000-Gate CMOS EPLD with Multiple Logic and Interconnect Arrays,* Proceedings. 1989 Custom Integrated Circuits Conference., May 1989, pp. 5.8.1-5.8.4.

Chapter 4
Erasable Programmable Logic Devices

The Technical Staff of Altera Corporation
Edited by Robert Hartmann

4.1. Introduction

Various acronyms are used to describe programmable logic, such as FPGA, GAL, PAL, EPLD, FPLA. Several of these acronyms have come to imply a particular architecture. EPLD, for example, is usually identified with array-based (AND-OR) programmable logic. This structure is particularly well-suited for implementing wide fan-in logic functions. These array-based structures were popularized in several families of bipolar, fuse-programmable logic devices but most particularly the PAL [Birkner 1978] which was introduced by MMI (Monolithic Memories, Inc. which was later acquired by Advanced Micro Devices.)

While these PAL devices were low in complexity by today's standards, they allowed the integration of several SSI and MSI components into one device. This integration advantage combined with high speed, flexibility to do both combinational and sequential functions, reasonable cost and the availability of software tools made these devices extremely popular. But, there were also drawbacks. The two most significant problems were high power dissipation and fuse programmability which meant that incorrectly programmed devices could not be salvaged. These disadvantages ultimately limited the logic complexity to a few hundred gates.

The EPLD (Erasable Programmable Logic Devices) developed by Altera overcame the limitations of bipolar fuse-programmed PALs by using CMOS technology in conjunction with floating-gate programmable transistors. This technology opened the way for reprogrammable circuits with much higher logic capacity and much lower power.

Altera's first architecture, known as Classic, was introduced in 1984 and offered chips ranging from 8 to 48 macrocells (see P. 7), which roughly translates to 300 to 2,000 gates [Hartmann 1984]. The Company's second architecture, introduced in 1988 and designated the MAX 5000 family, provides from 32 to 192 macrocells or 600 to 7,500 gates. In 1991 a third-generation architecture, called the MAX 7000 family, was introduced. The first member, the EPM7256, has 256 macrocells (approximately

10,000 gates). Devices with as many as 1,000 macrocells are planned. Each new generation has provided faster performance and greater gate density than preceding generations.

Proprietary software tools that make the use of EPLDs quick and simple were also developed. These software tools improve the productivity of the engineer, and can be used repeatedly by multiple users for different designs on an ongoing basis.

An EPLD logic design for a particular end use is achieved in three steps: design entry, design compilation, and design verification.

Design entry is accomplished using either the integrated software editors that are part of Altera's software, or -- if the customer prefers -- with tools provided by EDA vendors. Designs can be expressed as schematic diagrams, a text-based hardware description language, logic equations, state machines, truth tables, waveforms, or a combination of any of the above.

Design compilation of the design is performed by software that first "synthesizes" the logic specified during design entry so that the minimum number of logic cells (sometimes referred to as macrocells) are used. The software then arranges and interconnects these logic cells in the most compact manner possible (this is commonly called "fitting"). With early PLDs, logic synthesis was nothing more than factoring Boolean equations into the minimum number of product terms. If that number of generated product terms was less than or equal to the number available, then the "synthesis" was complete. If not, the user had to restructure the logic and try again.

Logic may be described in a variety of ways (gates, logic equations, hardware description language, etc.), all of which may be intermingled in the software environment. The job of logic synthesis is to take the logic and reduce it in such a way that it uses the minimum of a given device's resources. The synthesizer must have knowledge of the device's architecture and a set of minimization techniques in order to accomplish the task. In the Altera software, as many as eighteen minimization methods will be tried, with the result of the trials compared before the best implementation is selected. The proprietary algorithms that perform this function require minimal intervention from the user and offer highly-automated, "push-button" results.

Finally, in design verification, the designer confirms the logic functions performed by the EPLD and checks the timing of critical logic paths. Designs are verified using the static Timing Analyzer, Delay Predictor, and logic Simulator (which has detailed timing models for all of the internal EPLD elements), all of which are part of the design software tool kit.

4.2. Programming Technology

Prior to the introduction of the EPLD, the only technology used for Programmable Logic Devices (PLDs) was bipolar and fuse-based. The active elements on these devices were bipolar transistors with arrays of fuses providing programmable interconnect structures. These fuse elements consisted of a variety of exotic metal alloys and/or polysilicon structures, but all relied on the physical destruction of fuses to open connections by passing large currents through their small geometries.

The melting process in bipolar PLD fuses was difficult to control and often resulted in unacceptable programming yields. Since the process was irreversible, guaranteed results were impossible. The power-hungry bipolar technology also severely limited integration levels. With the advent of the EPLD, CMOS technology replaced bipolar technology, and fuses were replaced by reprogrammable EPROM or EEPROM transistors (EPLD bits). These EPLD bits were much smaller than fuses, electrically programmable, and erasable. Since the programming step was reversable, EPLDs were fully factory-tested, guaranteeing 100% programming yield at the customer site. The lower power required of CMOS technology allowed higher integration levels.

The EPROM cells operate via floating-gate charge injection [Wolf 1990]. The programming process consists of placing sufficient voltage (typically >12V) on the drain and gate of the transistor to create a strong electric field and energize electrons to jump from the drain region to the floating gate. Electrons attracted to the floating gate become trapped when the high voltage bias is removed. If the drain terminal is held at a low voltage during programming, electrons are not available to be attracted to the floating gate and the floating gate remains uncharged. Trapped charge changes the threshold of the EPROM cell from a relatively low value with no charge present ("erased") to a higher value when programmed. Figure 4.2.1 shows a basic cross-section of the EPROM cell technology.

The topographical view of a single EPROM cell is shown in Figure 4.2.2. The "Poly Select Line" in Figure 4.2.2 corresponds to the terminal "GATE" in Figure 4.2.1; the "Contact Area" and "Ground Diffusion" correspond to the "Drain" and "Source" regions of the EPROM transistor respectively.

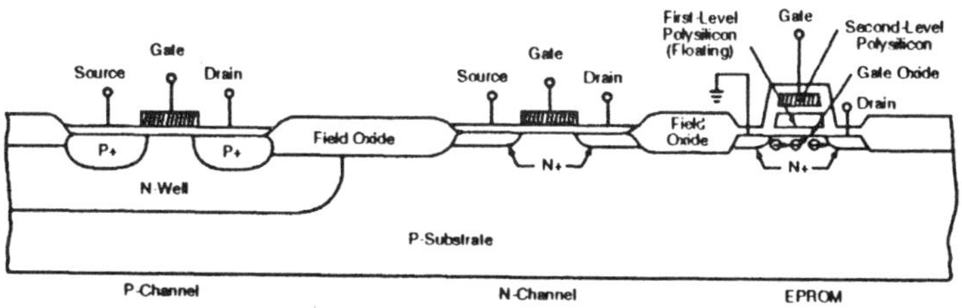

Figure 4.2.1 Cross Section of CMOS EPROM Die

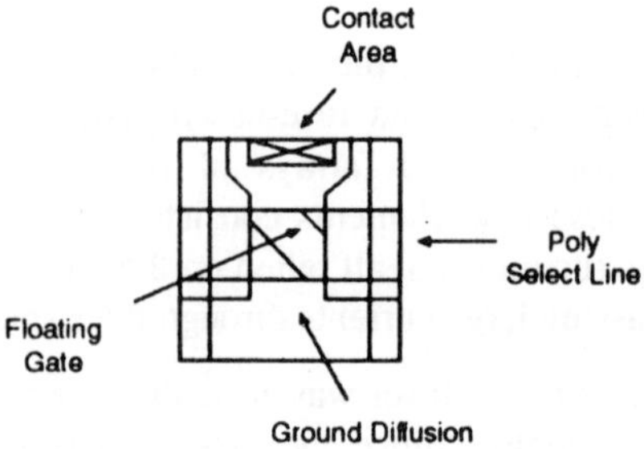

Figure 4.2.2 Top View of Single EPROM Transistor

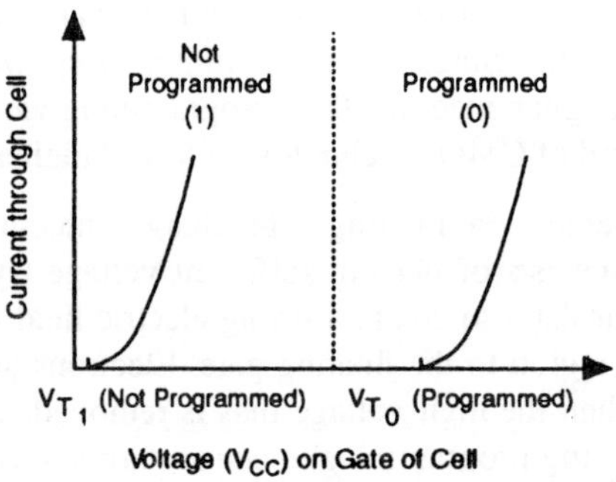

Figure 4.2.3 V-I Characteristic of EPROM Transistor

The EPROM transistor has two states, erased (unprogrammed) and programmed. The V-I characteristics for these two states is shown in Figure 4.2.3.

If it is assumed that the floating gate in the structure is initially unprogrammed, then the transistor works much like a normal N channel transistor. Whenever a positive voltage greater than the threshold voltage (VT1) is applied to the control gate, a channel is induced under the gate region which allows current to flow between the drain region and the source region. In a typical N channel floating gate device, VT is approximately 1.0 volts. The signals which are applied to the control gate are typically between 0 volts and 5 volts.

The effective threshold of the transistor can be changed as explained previously. After programming, charges remain trapped on the floating gate and cause the effective threshold of the EPROM transistor to be increased to a value greater than the voltage which would be applied to the control gate during Normal Operation. The application of the high voltages and the subsequent charge trapping on the floating gate is called programming.

After the transistor has been programmed, if a 5 volt signal is applied to the control gate, no channel will be created between the source and drain and no current will flow. This two layer transistor can be thought of as a programmable switch. In the unprogrammed state, the switch opens and closes in response to the application of 0 volts or 5 volts to the control gate. In the programmed state, the switch is always open regardless of whether 0 or 5 volts is applied to the control gate.

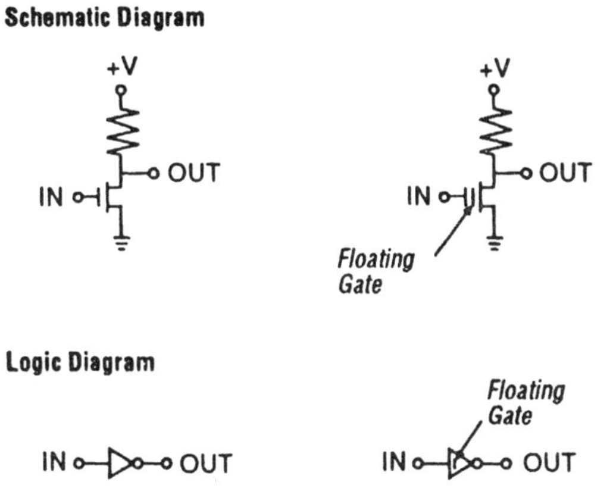

Truth Table

IN	OUT
+V	0
0	+V

(a)

IN1	Floating Gate	OUT
+V	Unprogrammed	0
0	Unprogrammed	+V
X	Programmed	+V

(X means either 0 or +V)

(b)

Figure 4.2.4 Use of an EPROM Transistor in a Programmable Inverter

Logic Structures Using EPROM Transistors

In the diagrams shown in Figure 4.2.4, an inverter is shown in part (a) using a standard N-channel transistor as the switching device and, in part (b) a floating gate transistor is shown as the switching element. Below each schematic diagram, a logic gate representation and the corresponding truth table is shown. The truth table for the EPROM inverter in the unprogrammed state, gives the same result as the normal N-channel inverter. However, when the floating gate is programmed, the output is always pulled high, independent of the input. Thus, a logic function (an inverter in this case) can be created wherein the input variable can be "programmed out" by programming the floating gate to which the input is connected.

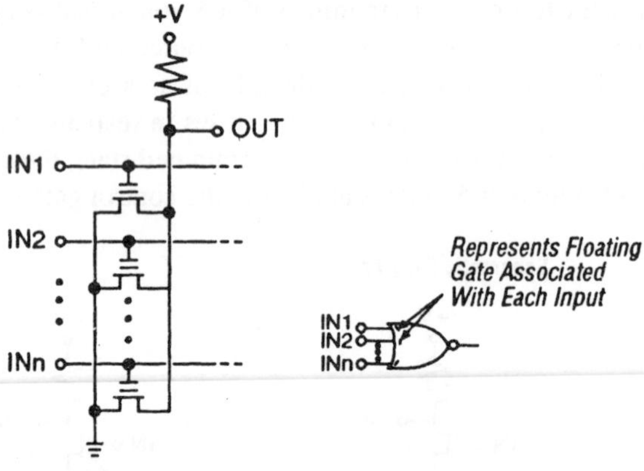

Schematic Diagram **Logic Diagram**

Truth Table

IN1	IN2	INn	OUT	
0	0	0	1	Note: P in the Truth Table means that the EPROM transistor at that site is programmed. X means either 1 or 0.
X	X	1	0	
X	1	X	0	
1	X	X	0	
X(P)	0	0	1	IN1 Gate is programmed
X(P)	1	X	0	
X(P)	X	1	0	
0	X(P)	0	1	IN2 Gate is programmed
1	X(P)	X	0	
X	X(P)	1	0	
0	0	X(P)	1	INn Gate is programmed
1	X	X(P)	0	
X	1	X(P)	0	

Figure 4.2.5 Programmable NOR Gate Using EPROM or EEPROM Transistors

In part (a) of Figure 4.2.5 a NOR gate circuit using EPROM elements is illustrated. In this circuit, the output is only a function of the inputs corresponding to floating gates left in the erased state. In part (b), the corresponding logic diagram is shown, and in part (c) the truth table for the circuit is shown.

This circuit is a one dimensional array of EPROM elements which forms a NOR gate, the elements of which are selectively programmable. By extending this concept to a two dimensional array and then collecting the programmable NOR outputs in another set of NOR gates, electrically programmable logic arrays such as illustrated in Figure 4.2.6 can be formed.

Any combinational logic function, however complex, can be expressed as a Boolean

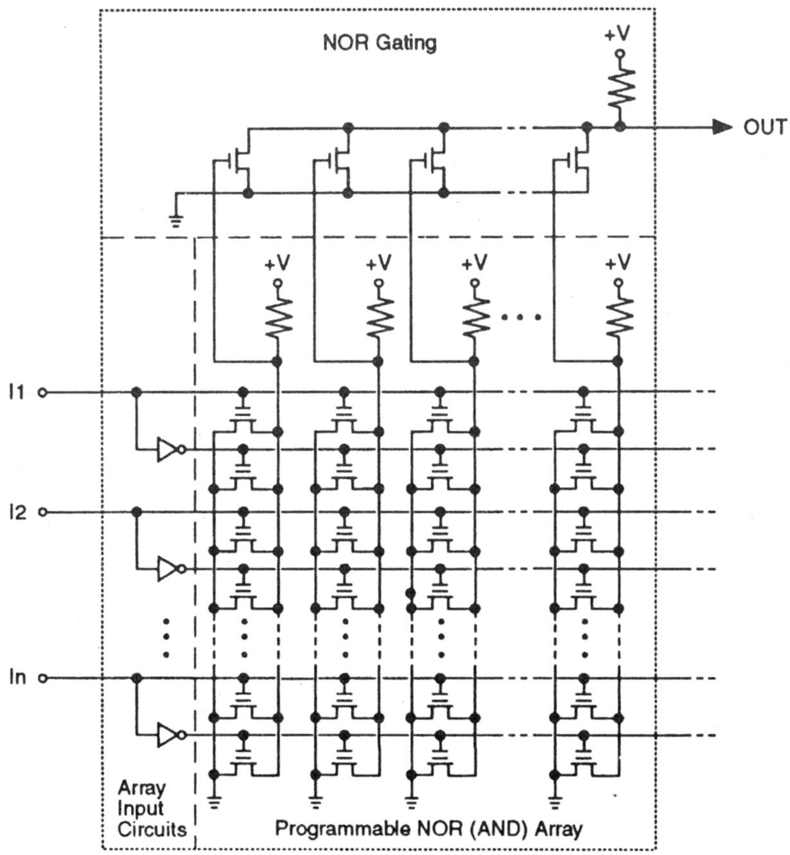

Figure 4.2.6 Programmable NOR Array

equation. Furthermore, any Boolean equation can be expressed in sum-of-products form. The logic expression is of the form f=P1+P2+P3+... where f is some Boolean variable and P1, P2, etc. are product terms. Product terms can be expressed as P = A1*A2*A3.... That is, product terms are the logical AND of literals. The hardware realization consists of an array (whose inputs are the true and complement of the literals) which generates the product terms and another array (or sometimes a simple OR gate) which generates the sum-of-product expression. Sequential functions are accommodated by adding a register (such as a D flip-flop). The input to the register is fed by the sum array output and the output of the register is fed back as an input to the product term array.

The NOR array shown in Figure 4.2.5 can be thought of as a product term generator and the sum-of-products can be generated by a structure like that shown in Figure

4.2.6. The connection from a particular input variable to a product term is a function of the state of the EPROM transistors.

Initially all of the EPROM transistors in the array are erased and are therefore connected to all of the product terms. By selectively programming these EPROM transistors, certain of the inputs are disconnected, leaving only those variables which are needed by a particular logic expression.

Each EPLD contains one or more AND arrays that provide product terms. A product term is simply an n-input AND gate, where n is the number of variables. EPLD schematics use a shorthand AND-array notation to represent large AND gates with common inputs. Figure 4.2.7 shows three different representations of the same logic function. Circuit A is presented in classic logic notation; Circuit B has been modified to a sum-of-products notation; and Circuit C is written in AND-array notation. A dot represents a connection between an input (vertical wire) and one of the 8-input AND gates. No dot implies no connection. Unused AND gate inputs float to a logic 1.

Even the rather simple 2X8 AND-array of Circuit C can produce all one and two variable functions, most three variable and some four variable functions. The table below shows the number of possible functions of one, two, three and four variables

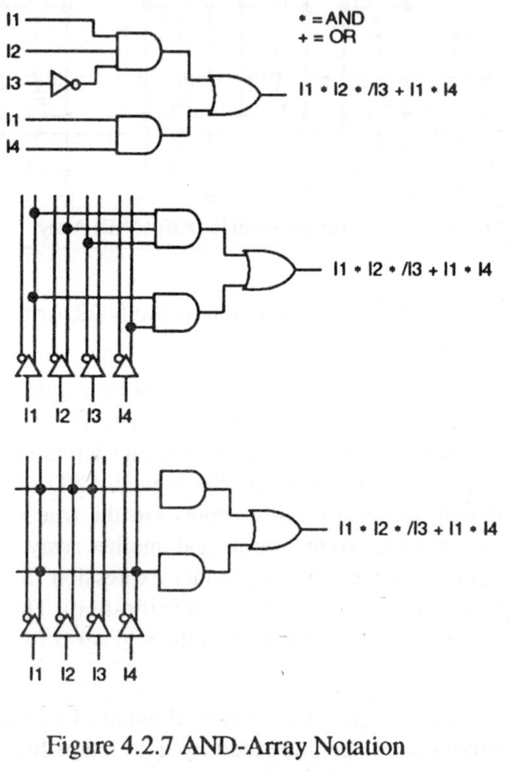

Figure 4.2.7 AND-Array Notation

and the ability of a two product-term AND-OR structure to synthesize these functions.

Number of Variables	Number of Functions Possible	Number of (1) Unique Functions	Coverage (2 product terms)
1	4	2	All
2	16	4	All
3	256	14	9 of 14
4	65,536	222	18 of 222

(1) Allows for inversions and reorderings of inputs and inversion of the output.

4.3. Device Architecture

Up to this point we have been discussing EPLDs at the level of transistors, EPROMs and product terms in order to provide the reader with some of the basic fundamentals. However, the use of software tools can eliminate the necessity of understanding any of the circuit-level complexities of EPLD architectures. The user may then work with familiar design entry tools (e.g., TTL macrofunctions or a high-level design language), and the software can automatically translate the design into the format required to fit the EPLD architecture.

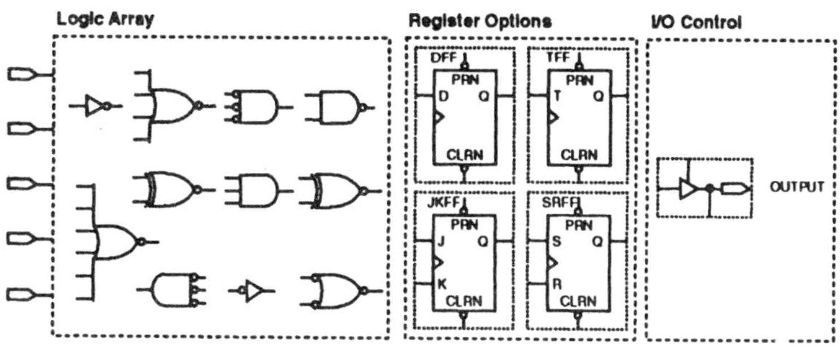

Figure 4.3.1 Basic Logic Resources of an EPLD

Basic Concepts

The heart of the EPLD is the logic array as discussed previously. In addition, general-purpose EPLDs contain a number of other resources, such as dedicated input pins, user-configurable I/O pins, and programmable flip-flop and clock options that maximize flexibility for integrating random logic functions. Figure 4.3.1 depicts a logic designer's view of the logic resources of a typical EPLD.

Input variables to the logic array come from the input and I/O pins and from the output of macrocells. The logic array outputs drive output pins and inputs to the flip-flops.

Macrocell Architecture

The fundamental logic building block of an EPLD is the macrocell. Each macrocell consists of three parts:

- The logic array which implements all combinational logic functions.
- The programmable register which can be configured to provide D, T, JK, or SR options (the register can also be bypassed).
- Programmable I/O. Each I/O pin can be configured as a dedicated output, a dedicated input, or as a bidirectional pin.

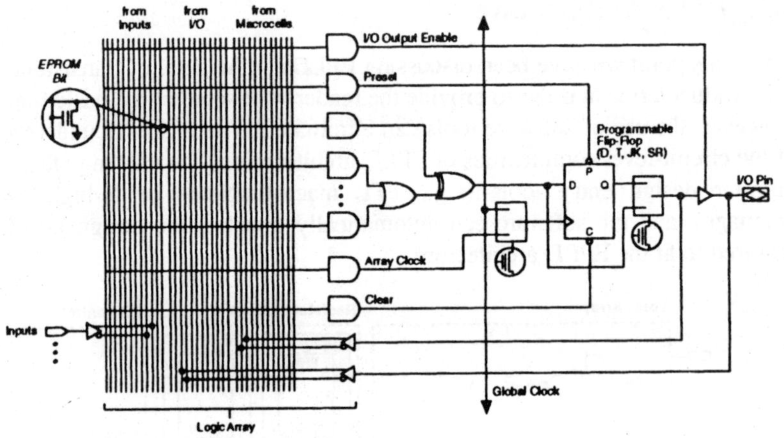

Figure 4.3.2 Typical EPLD Macrocell Logic Diagram

Logic Array

Figure 4.3.2 is a diagram of a typical EPLD macrocell.

The logic array is a programmable-AND/fixed-OR programmable logic array. The inputs to the AND array come from dedicated input pins, I/O pins, and from macrocell feedback paths. Typically, both true and complement versions of any of these signals are available as inputs to the programmable AND gates.

The macrocell shown in Figure 4.3.2 contains product terms that can be used for both combinational and sequential functions. Connections between the array inputs and the product terms are created during the programming process. One may think of a product term as an AND gate with many possible true/complement input signal pairs. Any product term may be connected to the true or complement (or both or neither) of any array input signal. If both the true and complement of any input signal are left intact, a logical false results on the output of the product term. If both the true and complement connections of any input signal are open, a logical "don't care" results for that input. If all inputs for the product term are programmed open, a logical true results on the output of the product term.

Several product terms feed a fixed OR whose output connects to an exclusive-OR (XOR) gate. The second input to the XOR function is controlled by a programmable resource (usually a product term) that allows the logic array output to be inverted. Software products which support this architecture can take advantage of this XOR function to implement active-high or active-low logic, complex mutually exclusive and arithmetic functions, or to reduce the number of product terms needed to implement a function (by applying De Morgan's inversion). Figure 4.3.3 illustrates an example of an OR function that requires six product terms in its original form. By using the "programmable" XOR gate and De Morgan's inversion, the OR function can be transformed into a NAND function:

A+B+C+D+E+F = / (/A*/B*/C*/D*/E*/F)

This inversion from OR to AND allows the equation to be implemented in a single product term.

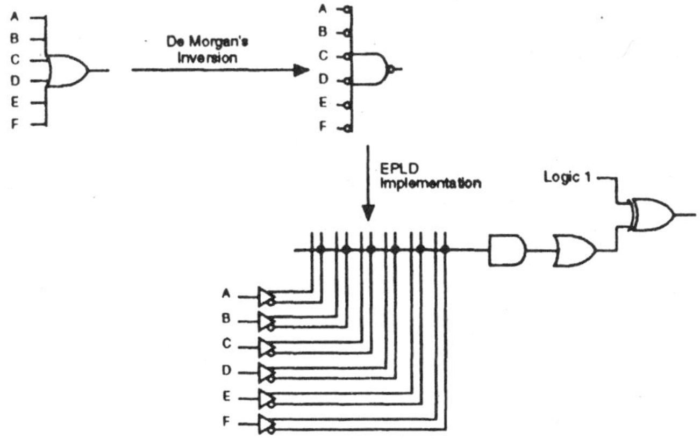

Figure 4.3.3 Product Term Reduction Using De Morgan's Inversion

Programmable Flip-Flops

Programmable flip-flops are used to create a variety of logic functions that use a minimum of EPLD resources. Each flip-flop can be programmed to provide a conventional D, T, JK, or SR function. In some devices, the macrocell flip-flops can also be configured as flow-through latches. Macrocell flip-flops may also have an asynchronous Clear and Preset capability which allows complete emulation of many commonly used TTL macrofunctions.

Programmable Clock

In many PLD architectures, the clock source for the macrocell flip-flop is programmable. Each flip-flop may be clocked from a dedicated device input pin (also known as a synchronous or global clock), or from any input or I/O pin, or any internal logic function (via a product term). The product term clock source is often called the array clock. A clock selection circuit associated with each flip-flop is programmed to make the desired choice. Flip-flops can thus be clocked independently or in user-defined groups. Macrocell flip-flops are typically positive-edge-triggered with data transitions that occur on the rising edge of the clock. Array clocks allow positive or negative edge triggered clocks, gated-clocks and clock-enable logic to be implemented. However, global clock signals have faster clock-to-output delay times than internally-generated product term clock signals.

I/O Control Block

Figure 4.3.4 illustrates the resources associated with a typical I/O pin. The I/O block contains a tri-state buffer whose data input comes from a macrocell. The tri-state function is controlled by a macrocell product term. I/O pins may be configured as dedicated outputs, as bidirectional outputs, or as additional dedicated inputs. In most PAL architectures (PAL 16R8 or 22V10, for example) there is a physical coupling of I/O pins to macrocells. In such architectures, when an I/O pin is used as an input, the macrocell associated with that pin cannot be used because the macrocell feedback path is used by the I/O input signal. This is a waste of a valuable resource. More recent architecture (e.g., MAX 5000 and MAX 7000) have "dual feedback", whereby the macrocell feedback is decoupled from the I/O pin feedback. Dual feedback makes it possible to implement a "buried" function in the macrocell while the I/O pin is used simultaneously as a dedicated input. Applications that require many buried flip-flops such as counters, shift registers, and state machines, or bus-oriented functions are more easily accommodated by this type of programmable I/O block.

Design Security

All EPLDs contain a programmable Security Bit that controls access to the data programmed into the device. If this feature is used, a proprietary design implemented in the device cannot be copied or retrieved. This feature provides a high level of design security, since programmed data within EPROM or EEPROM cells is invisible. The Security Bit that controls this function, as well as all other program data, is reset by erasing the EPLD.

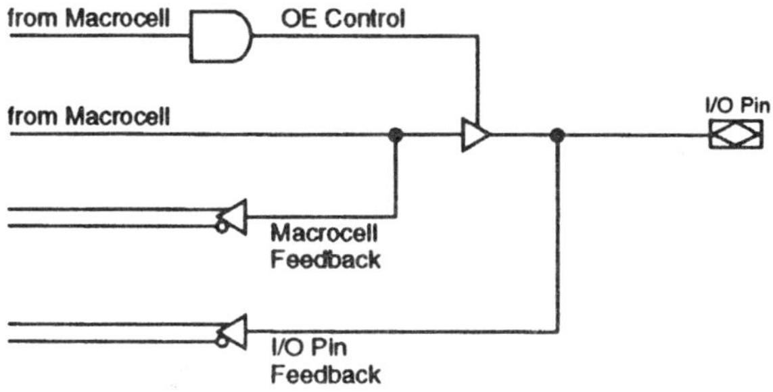

Figure 4.3.4 EPLD I/O Pin with Dual Feedback

Functional Testing

Because they are erasable and reprogrammable, EPLDs may be fully tested, and conformance to specification may be guaranteed. Complete testing of each programmable EPLD bit and all internal logic elements ensures 100% programming yield. Test programs may be used and then erased during early stages of the production flow. This facility to use application-independent, general-purpose tests, called generic testing, is unique to reprogrammable user-configurable logic devices. EPLDs also contain on-board logic test circuitry to allow verification of function and AC specifications of devices in windowless packages.

Operating Requirements for EPLDs

Most devices operate at a nominal power supply voltage of 5 volts (some 3.3 volt parts are also available). Input levels and output drive characteristics are consistent with TTL family characteristics.

Certain precautions are required for trouble-free operation. Unused inputs must be tied to VCC or GND. Unused I/O pins should also be tied to VCC or GND. Each set of VCC and GND pins must be decoupled directly at the device.

Whenever many output pins are switching simultaneously, system noise can be generated. This noise is usually seen on the Vcc supply or on adjacent quiescent outputs. In severe cases, this noise can cause unwanted triggering of flip-flops, either on the device itself or on other devices in the system. Evolving circuit design techniques have done much to eliminate the problem. However, faster slew rate outputs and higher drive capability of new devices make the circuit design problem ever more difficult. As a general precaution whenever a design allows eight or more outputs to switch simultaneously, special precautions in PC board design are recommended. Among these are: use of embedded Vcc and GND planes, scrupulous use of decoupling capacitors, restriction of signal trace length to eight inches or less, and use of small series resistors (10 ohms to 30 ohms) on long traces or those with highly capacitive loads.

Architectural Evolution in Array-Based PLDs

In the preceding sections the general features of array-based architectures have been discussed. The evolution began with the basic programmable AND array whose outputs feed either a fixed OR (the PAL structure) or a programmable OR (the FPLA structure) to produce sum-of-products logic implementation. Flip-flops were added to this structure to allow implementation of sequential functions. Many ancillary functions were then added to provide for better utilization of the parts. More flexible I/O structures, programmable flip-flop clock control, programmable flip-flop types (e.g., D, J-K, T) are examples of these enhancements. However, these early parts beginning with the PAL (MMI) and continuing through the 22V10 (AMD), the EPxxx series (Altera) and the GAL (Lattice) all were single array, globally connected parts. "Globally connected" means that the output signal(s) of every macrocell is fed back as an input to the logic array.

A fundamental problem with simple expansion of macrocell-based PLDs is that as the number of macrocells grows by N, the programmable elements (e.g., EPROM or EEPROM bits) in the array which feeds those macrocells grows by N-squared [Wong 1989]. There are two problems with increasing the bit count: first, the yield suffers because of the larger die size, and second, the delay through the array increases. Much of the architectural work that has gone on in array-based PLDs has been directed at breaking the N-squared array growth relationship to increasing complexity. It has been a primary focus of research at Altera, AMD, Lattice and others. In the sections that follow, we will explore the architectural evolution using three generations of products developed at Altera. As we look at each product, we will try to point out the problems that existed and the architectural solution that was implemented. While this treatment is not exhaustive, the discussion of these product architectures will provide a good basis for understanding the product architectures of other vendors.

4.3.1 - The "Classic" Family of PLDs

The original family of Altera EPLDs are grouped into what is known as the "Classic" family. The largest member of this family is the EP1810. The architecture used in the EP1810 is an early attempt at breaking single, large, globally-connected arrays into a number of smaller "local" arrays with a connection structure between these local arrays. An understanding of the EP1810 architecture provides a good foundation for understanding the more complex MAX 5000 and MAX 7000 products to be described in a later section.

Functional Description of the EP1810

Figure 4.3.1.1 shows the complete block diagram of an EP1810 EPLD. The EP1810 device has four identical quadrants, each containing 12 macrocells. Internal bus structures in these EPLDs feed input signals into the macrocells. Macrocell outputs drive the external pins and internal buses.

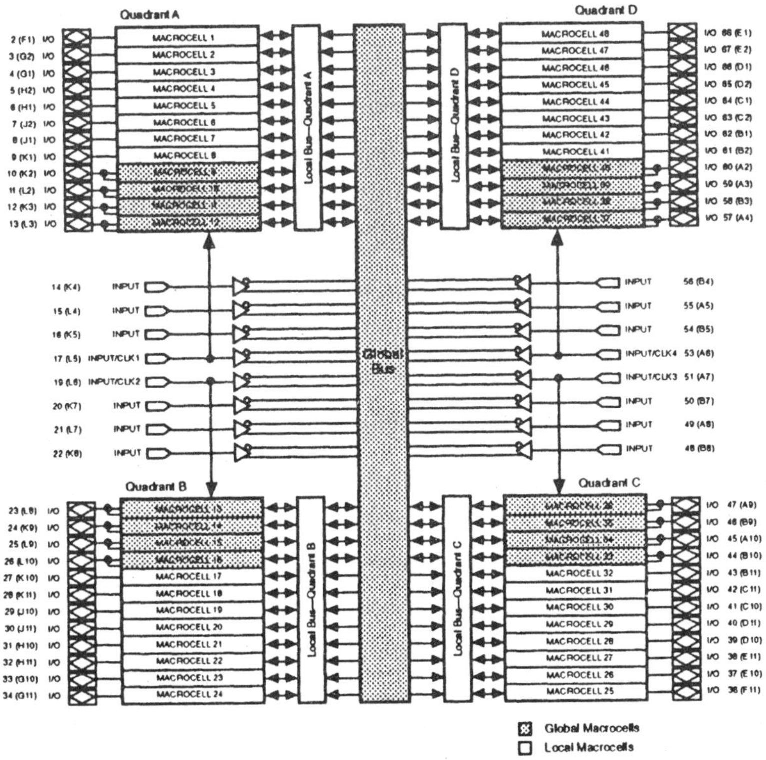

Figure 4.3.1.1 EP1810 Block Diagram

EP1810 EPLDs use CMOS EPROM cells to configure internal logic functions. The architecture is 100% user-configurable, accommodating a variety of independent logic functions. Figures 4.3.1.2 and 4.3.1.3 depict the detailed macrocell architecture used in this device.

EP1810 EPLDs have 48 macrocells, 16 dedicated data inputs, 4 global Clock inputs, and 48 I/O pins that can be individually configured for input, output, or bidirectional operation on a macrocell-by-macrocell basis. Each macrocell contains 10 product terms for the following functions: 8 product terms are dedicated to logic implementation; 1 product term is used for Clear control of the internal register; and 1 product term implements either Output Enable or an array Clock.

Of the 48 macrocells, 32 are local (see Figure 4.3.1.2) and 16 are global macrocells (see Figure 4.3.1.3). Local macrocells offer a multiplexed feedback path (with pin or macrocell feedback) and drive the local bus in their quadrant. Global macrocells feature two dedicated feedback paths: one feeds the local bus; the other feeds the global bus. This arrangement, called "dual feedback," allows global macrocells to

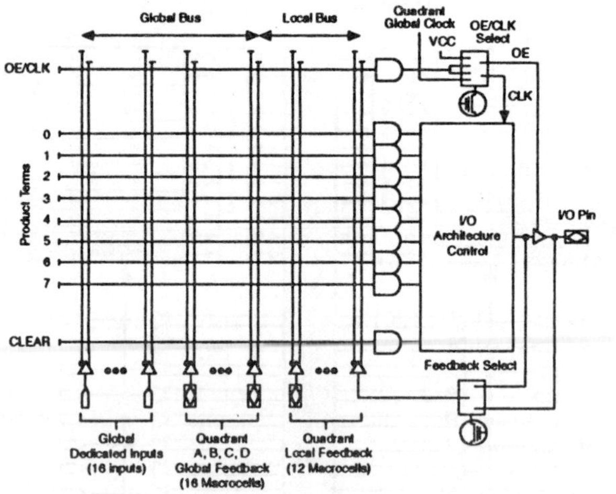

Figure 4.3.1.2 EP1810 Local Macrocell

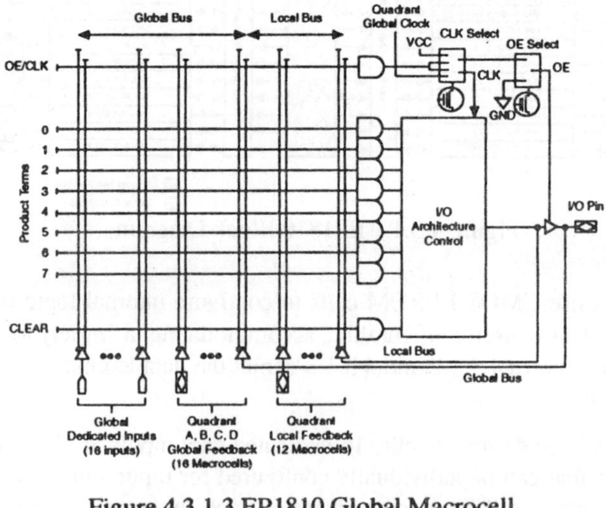

Figure 4.3.1.3 EP1810 Global Macrocell

implement buried logic functions while the associated I/O pin is used as an input. Dual feedback ensures maximum I/O flexibility.

Each macrocell consists of a logic array, a tri-state I/O buffer, and a selectable register element that can be programmed for D, T, J/K, or S/R operation, or bypassed for combinational functions. Each macrocell also has programmable output polarity. The logic array has a sum-of-products (AND/OR) structure. The 88 inputs to the programmable-AND array come from true and complement signals of the 16 dedicated data inputs, the 12 local feedback signals, and the 16 global feedback signals. The EP1810 EPLD has a total of 480 product terms distributed among 48 macrocells. Each product term represents an 88-input AND gate.

Clock Options

Each internal flip-flop in EP1810 EPLDs can be clocked independently or in user-defined groups. Each internal register may select its clock source from a dedicated global clock pin or a product term within the macrocell. Any input or internal logic function can be used as a clock.

Four dedicated global clocks (CLK1 to CLK4) also provide global clock signals to the flip-flops. One global clock is located in each quadrant; each of which is connected directly to an EP1810 external pin. Global clocks provide clock-to-output delay times that are faster than internally generated clock signals. Array clocks provide individual clocking on a macrocell-by-macrocell basis, either directly from pins or through internal logic. Array clock signals allow flip-flops to be configured for positive- or negative-edge-triggered operation. When global clocks are used, the flip-flops are triggered by the positive edge, i.e., data transitions occur on the rising edge of the clock.

Segmentation

The EP1810 was constructed using 4 quadrants, each containing 12 macrocells. These "clusters" of macrocells are called Logic Array Blocks, or LABs.

N^2 growth in the EP1810 was curtailed by going to a four quadrant (LAB) architecture. This arrangement provided local feedback to each LAB from the 12 local macrocells and global connectivity to the other LABs from 4 of the 12 macrocells in each LAB.

This architectural concept is effective because logic can be partitioned into functions whose internal connectivity is high but whose external connectivity to other functional units is limited. The intra-connectivity of signals in the LAB is very rich (every macrocell output feeds every macrocell within the LAB), while the inter-connectivity between LABs is more limited (restricted to signals generated by the global macrocells). The concept of the LAB for local functions and global lines which allow communication between modules was significantly improved in the MAX 5000 series which will be discussed next.

4.3.2 - The MAX (Multiple Array matriX) Product Family

General Description

The MAX 5000 family of EPLDs represent a significant step in the architectural evolution of programmable logic. Improvements in architecture, process technology and design result in significant increases in logic density, flexibility, and speed. The largest member of the family, the EPM5192, replaces over 100 7400-series SSI and

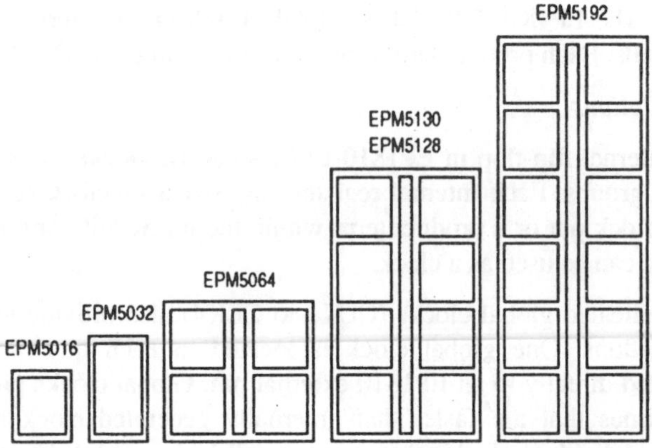

	EPM5016	EPM5032	EPM5064	EPM5128/EPM5130	EPM5192
Macrocells	16	32	64	128	192
Maximum Flip-Flops	21	42	84	168	252
Maximum Latches	32	64	128	256	384
Pins	20	28	44	100/84/68	100/84

Figure 4.3.2.1 MAX 5000 Family

MSI packages, allowing integration of complete subsystems into a single package, saving board area, and reducing power consumption.

The MAX 5000 EPLDs range in density from 16 to 192 macrocells (see Figure 4.3.2.1). They are divided into two groups: high-speed EPLDs (EPM5016 and EPM5032) and high-density EPLDs (EPM5064, EPM5128, EPM5130, and EPM5192). The high-speed devices achieve system clock frequencies of 66 MHz, and are capable of attaining counter (of up to 32 bits) frequencies of 100 MHz, while the high-density parts achieve system clock frequencies of 35 MHz and counter frequencies of 62.5 MHz.

The modular architecture of MAX 5000 EPLDs provides integration solutions over a wide range of logic densities. Because of the uniformity of the architecture, migration from one type of device to another is easy. For example, the EPM5128 and EPM5130 EPLDs have the same logic capacity, but have packages optimized to handle different I/O requirements. The EPM5128 comes in a 68-lead package, has eight dedicated inputs and 52 I/O pins, while the EPM5130 comes in a 100-lead package and has 20 dedicated inputs and 64 I/O pins. Over the entire family, a wide range of packaging options for both through-hole and surface-mount applications is available.

Logic Array Block

The EPM5016 and EPM5032 EPLDs have a single Logic Array Block (LAB). The EPM5064, EPM5128, EPM5130, and EPM5192 EPLDs contain multiple LABs. Each LAB contains a macrocell array, an expander product term array, and a decoupled I/O block (Figure 4.3.2.2). Expander product terms (expanders) are unallocated, inverted product terms that can be used and shared by all macrocells in the LAB. In the higher-density devices (EPM5064 and larger), macrocell output signals are routed between multiple LABs by a Programmable Interconnect Array (PIA) that ensures 100% routability. This multiple array architecture enables MAX 5000 EPLDs to offer the speed of smaller arrays (fMAX up to 62.5 MHz within the LAB of a EPM5192) with the integration density of larger arrays.

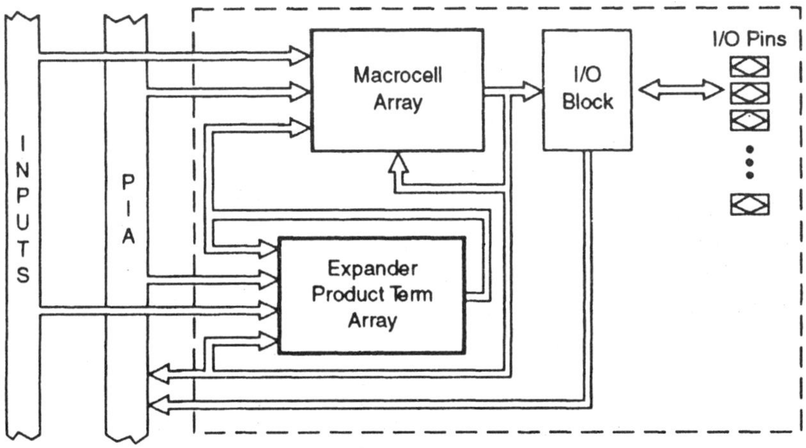

Figure 4.3.2.2 MAX 5000 Logic Array Block

MAX 5000 EPLDs contain from 1 to 12 Logic Array Blocks (LABs). Macrocells are the primary resource for logic implementation, but if needed, expanders can be used to supplement the capabilities of any macrocell. Flexible macrocells and allocatable expanders facilitate variable product term designs without the waste associated with fixed product term architectures. The outputs of the macrocells feed the decoupled I/O block, which consists of a group of programmable tri-state buffers and I/O pins. In the EPM5064, EPM5128, EPM5130, and EPM5192 EPLDs, multiple LABs are connected by a Programmable Interconnect Array (PIA). All macrocell outputs are globally routed within a LAB and also feed the PIA to provide efficient routing of signal-intensive designs.

Macrocells

The MAX 5000 macrocell, shown in Figure 4.3.2.3, consists of a programmable logic array and an independently configurable register. This register may be programmed for D, T, JK, or SR operation; or as a flow-through latch; or bypassed for purely combinational operation. Combinational logic is implemented in the programmable logic array, which consists of three product terms ORed together that feed one input of an XOR gate. The second input to the XOR gate is also controlled by a product term. The output of the XOR gate feeds the programmable register. Expanders can be allocated to enhance the capability of the logic array.

Additional product terms, called secondary product terms, are used for output enable, preset, clear, and clock logic. Preset and clear product terms drive the active-low asynchronous preset and asynchronous clear inputs to the configurable flip-flop. The clock product term allows each register to have an independent clock and supports positive- and negative-edge-triggered operation. Macrocells that drive an output pin may use the output enable product term to control the active-high tri-state buffer in the I/O control block. These secondary product terms allow 7400-series TTL functions to be emulated exactly.

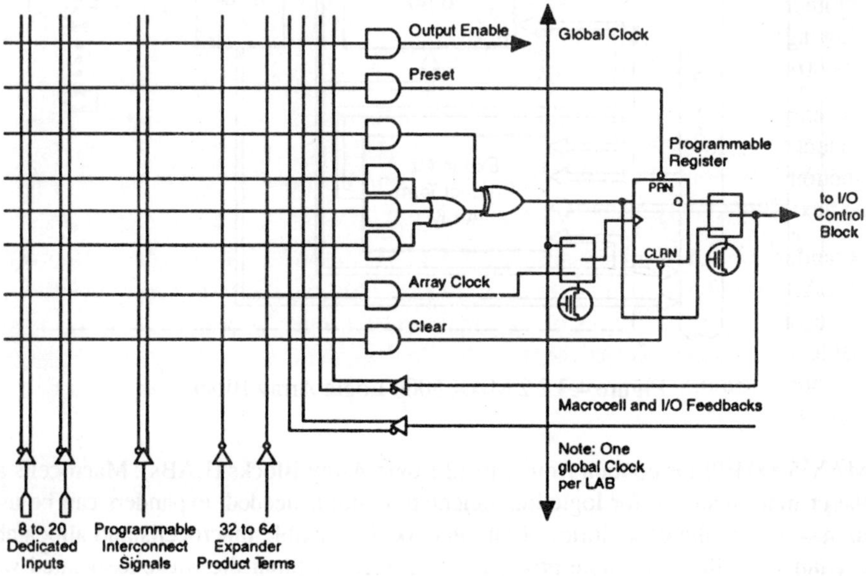

Figure 4.3.2.3 MAX 5000 Macrocell

Expander product terms

In programmable AND, fixed OR structures the chip architect must decide how many product terms is "enough" -- a typical choice has been eight. In some architectures [Kitson 1984] up to 16 product terms feed some OR gates. In most applications, eight product terms are more than enough, but are insufficient in some particular cases. This is consistent with the data published by Munoz [1987] which indicates that approximately 70% of all combinatorial functions require three (or less) product terms. Further, a substantial minority of functions require more than eight with the tail of the distribution exceeding 16.

This presents a significant dilemma for the chip architect. The choice of a high number of product terms per OR (e.g., eight or higher) will result in a larger, slower chip where many of the resources will typically be wasted. The choice of a low number of product terms per OR (e.g., three) can result in difficulty implementing more complex functions.

The MAX 5000 architecture has three product terms per OR gate, which then feeds one input of an XOR gate, plus a fourth product term which feeds the other input of the XOR gate.

This allows implementation of all three product term functions and many four product term functions in a single pass through the array. Expanders were added to accommodate more complex logic expressions. When expanders are used the architecture becomes more like a programmable-AND, programmable-OR structure, allowing implementation of the most complex functions. However, an additional delay (the expander delay) is incurred.

The expander product term array (Figure 4.3.2.4) can be used and shared by all product terms in the LAB. Wherever extra logic is needed (including register control functions), expanders can be used to implement the logic. These expanders provide the flexibility to implement both register and product-term-intensive designs.

Expanders are fed by all signals in the LAB. One expander may feed all macrocells in the LAB or multiple product terms in the same macrocell. Since expanders also feed the secondary product terms of each macrocell, complex register control and output enable logic functions can be implemented without using additional macrocells. Expanders can also be cross-coupled to build additional flip-flops or latches [Altera AB76 1990].

Clock Options

Each LAB has two clocking modes: array and global. If array mode is chosen, each flip-flop is clocked by a product term. Thus, any input or internal logic function may be used as a clock. This allows systems that require multiple clocks to be easily integrated.

Global clocking is provided by a dedicated clock signal (CLK) from a single device pin. This direct connection provides shorter clock-to-output delay times. Each LAB has one global clock line which can be connected (programmably) to the global clock signal. If connected, all flip-flops within the LAB are positive-edge-triggered from the CLK pin. If the global clock is not connected to a particular LAB, then the flip-flops within the LAB are clocked by the individual array clock product terms associated with each flip-flop. If the CLK pin is not used as a global clock, it may be used as a dedicated input.

I/O Control Block

Each LAB has an I/O control block (Figure 4.3.2.5) that consists of a user-configurable I/O control function for each I/O pin. The I/O control block is fed by the

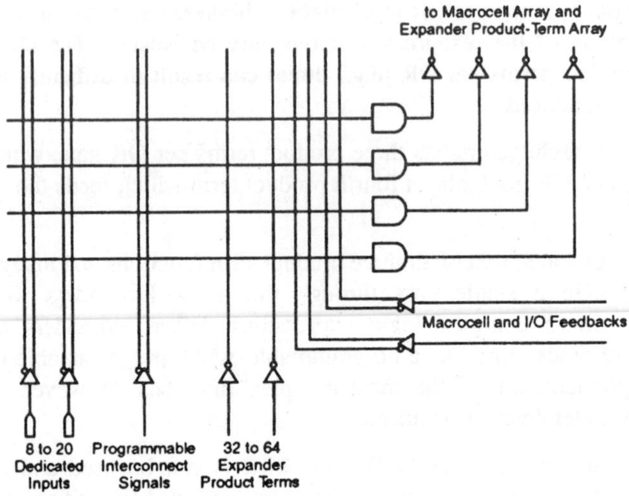

Figure 4.3.2.4 MAX 5000 Expander Product Terms

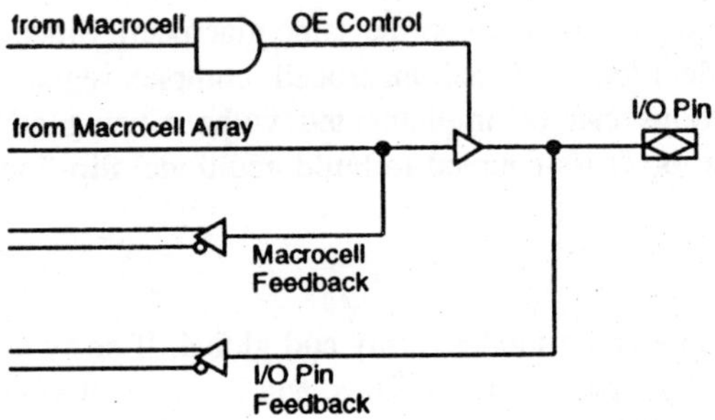

Figure 4.3.2.5 MAX 5000 I/O Control Block

macrocell array. The tri-state buffer is controlled by a dedicated macrocell product term, and drives the I/O pad.

Dual feedback -- a feedback path both before and after the tri-state buffer -- is employed for every I/O pin. This structure effectively decouples the I/O pins from the macrocells so that all macrocells within the LAB can be "buried." Thus, I/O pins can be configured as dedicated input, output, or bidirectional pins. In multi-LAB devices, I/O pins feed the PIA.

Programmable Interconnect Array

The higher-density MAX 5000 devices (EPM5064, EPM5128, EPM5130, and

EPM5192) use a Programmable Interconnect Array (PIA) to route signals between the various LABs. The PIA in the MAX 5000 devices is a fully populated cross-point switch. All I/O inputs and all macrocell outputs are inputs to the switch. There are 24 output lines per LAB. Switch connections are made by programming EPLD bits at the intersection of each unwanted crosspoint. With this PIA structure, any macrocell output signal, or any I/O input signal, may be routed to any other macrocell input without the chance of being blocked. The only routing limitation is the 24 signal fan-in limitation. The PIA has a fixed, path independent, delay, which makes timing performance easy to predict.

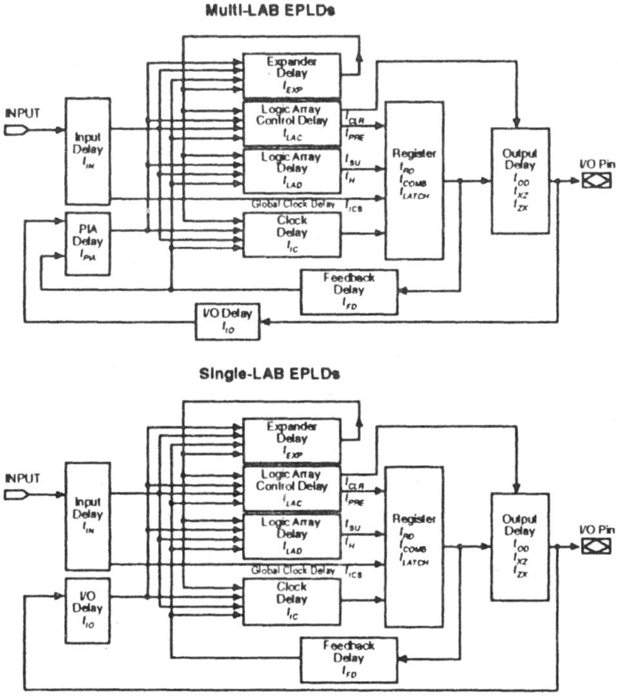

Figure 4.3.2.6 MAX 5000 Timing Model

Timing Model

Timing within multi-array parts can be determined in software, either with a Static Timing Analyzer tool, or with models as shown in Figure 4.3.2.6. Fixed internal delays allow the user to determine the worst-case timing delays for any design.

The timing models shown in Figure 4.3.2.6 may be used together with the internal timing parameters for a particular EPLD to derive timing information. External timing parameters are derived from a sum of internal parameters and represent pin-to-pin timing delays. Figure 4.3.2.7 shows the internal timing waveforms for these

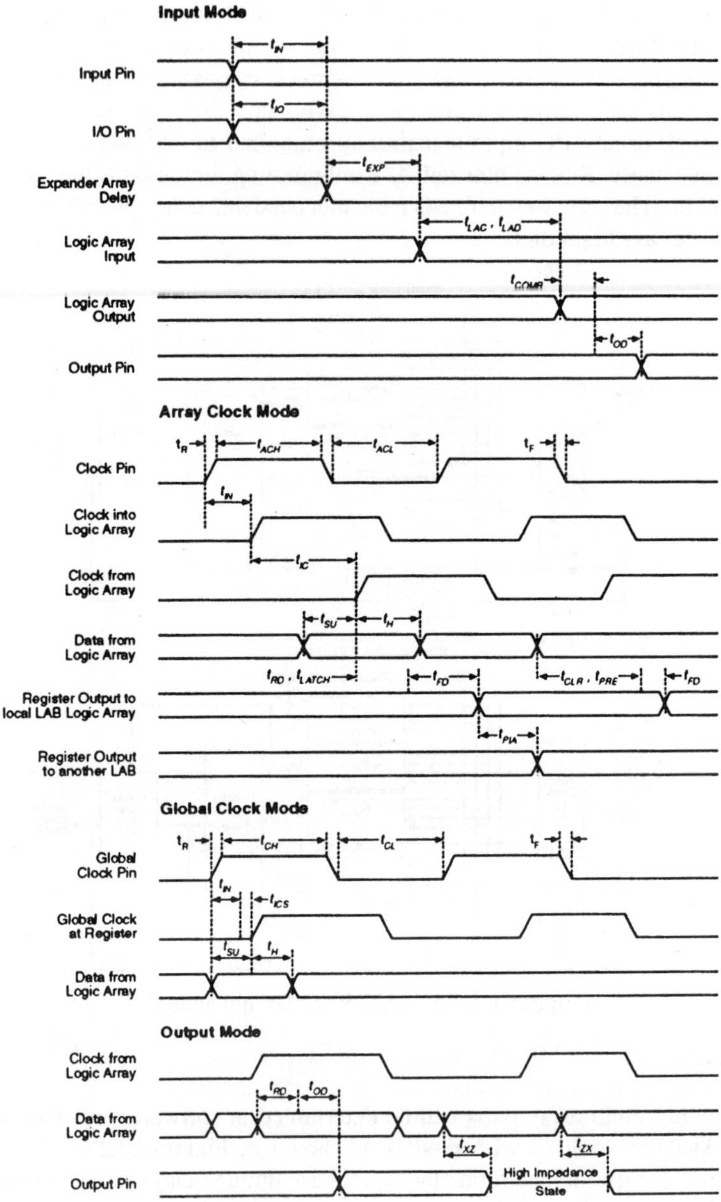

Figure 4.3.2.7 MAX 5000 Switching Waveforms

devices. Refer to Section 4.6 for further information.

4.3.3 - MAX 7000

General Description

The MAX 7000 family of EPLDs is a third generation architecture from Altera. The major goals of the MAX 7000 family are higher speed, lower power, lower cost, greater logic capability and more I/O for a given logic density. EEPROM technology is employed as the programmable element, allowing reprogrammability in non-windowed plastic packages.

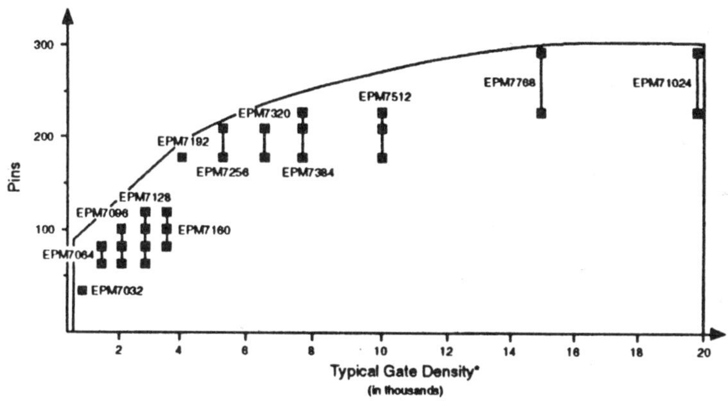

Figure 4.3.3.1 MAX 7000 Pin Count Versus Gate Density

MAX 7000 EPLDs use a multi-array style architecture to build devices with logic densities up to 10,000 gates. This family of EPLDs (shown in Figure 4.3.3.1) supports 10-ns pin-to-pin logic delays and 100-MHz system clock frequencies. Devices are housed in J-lead chip carrier (JLCC and PLCC), pin-grid array (PGA), and quad flat pack (QFP) packages, providing 44 to 288 pins.

MAX 7000 EPLDs are the first programmable logic devices with programmable speed/power optimization. Speed-critical sections of the design can run at high speed, requiring full power, while the non-speed critical sections can run at reduced speed while dissipating less than one-half of the power of full-speed operation.

The multiple-array architectural concept, pioneered in the MAX 5000 family, has been studied extensively at Altera. For example, LABs of 4, 8, 16 and 32 macrocells, both with and without expanders, have been considered. Fan-in to the LABs from the PIA was also varied. Each architecture variant was tested using a data base of over 100 design files and evaluated for utility (how many designs fit), speed, and cost (die size).

The MAX 7000 devices build on the information learned from this research. While the basic architectural choices such as LAB logic capability (e.g., 16 macrocells); global routing within the LAB; and a programmable interconnect array are carried over to the MAX 7000 family, some significant changes have been made. The key architectural changes are described below.

- A re-vamped PIA structure which provides significantly faster routing in the PIA.
- Product term selection matrix in the LAB to more efficiently use the product term resources.
- Ability to accommodate wide input OR functions (called parallel expanders) by borrowing unused product terms from adjacent macrocells.
- Programmable power saver mode. Each macrocell can be individually programmed to dissipate less power if lower speed performance is acceptable.
- Substantial increase in pin-to-logic ratio in order to accommodate I/O-intensive data path applications and 32-bit microprocessor support logic.

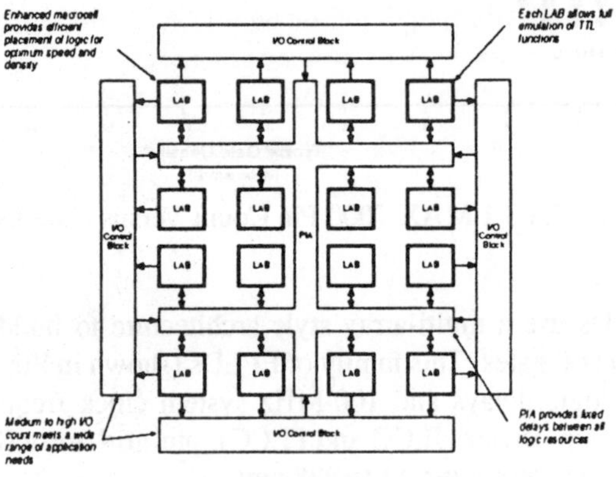

Figure 4.3.3.2 MAX 7000 Block Diagram

Functional Description

The MAX 7000 architecture (shown in Fig. 4.3.3.2) includes the following five basic elements:

- Logic Array Blocks
- Macrocells
- Logic expanders (shared and parallel)
- Enhanced programmable interconnect array
- I/O control blocks

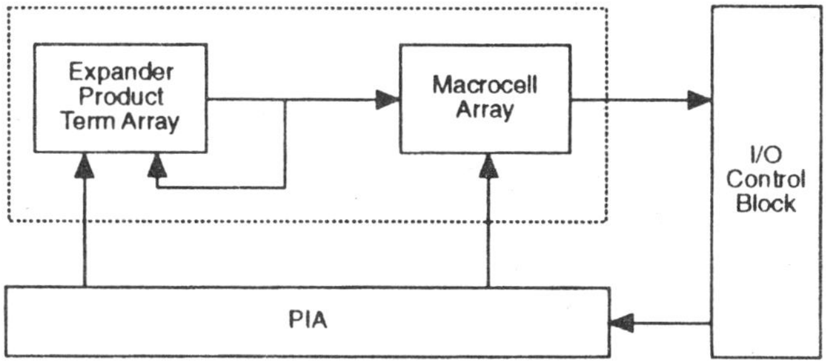

Figure 4.3.3.3 MAX 7000 Logic Array Block

Logic Array Blocks

The MAX 7000 Logic Array Block is shown in Figure 4.3.3.3. Each LAB contains 16 macrocells, and up to 16 shared logic expanders. Shared expanders can provide additional logic resources to any of the macrocells in an LAB. Furthermore, since all macrocells within an LAB share logic inputs, if one macrocell uses a specific logic input or a shared logic expander, it is also available to all other macrocells within that LAB.

Each LAB is fed by 36 inputs from the PIA, providing sufficient fan-in for the 16 macrocells to implement a wide range of typical logic functions. If more inputs are needed (for example in very wide data paths) several LABs can be used in parallel.

Macrocells

Macrocells within the LAB provide both sequential and combinational logic capability, thus ensuring the most efficient implementation of a wide range of logic functions. The MAX 7000 macrocell is shown in Figure 4.3.3.4.

Each macrocell has one flip-flop that can be programmed for D, T, JK, or SR operation with programmable clock control, individually configured for each macrocell. If necessary, the flip-flops can be bypassed for combinational operation.

Along with a programmable flip-flop, each macrocell also contains five basic product terms. These product terms can be allocated by a Product-Term-Select Matrix as primary inputs for combinational functions; as secondary inputs for individual Clear, Preset, Clock, and Clock Enable logic functions for the flip-flops; or as logic expanders to assist the generation of complex logic functions.

In addition, global Clock, Clear, and Output Enable control signals come in directly from device pins, eliminating the logic array delay and minimizing control-function delays.

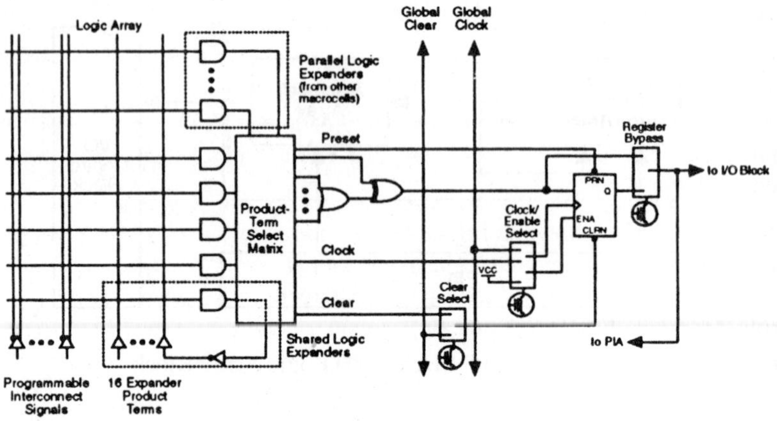

Figure 4.3.3.4 MAX 7000 Macrocell

The Clock Enable function allows flip-flops to be controlled by the logic array, even when they are clocked from the fast global Clock. This feature facilitates the implementation of high-speed synchronous designs. The Clock Enable function also allows each macrocell register to be clocked individually.

Logic Expanders

Whereas most logic can be implemented with the five basic product terms in each macrocell, some logic functions are more complex and require more product terms. Instead of using another macrocell to supply the additional logic resources, expanders are available to provide additional product terms directly to any macrocell. Unlike MAX 5000 EPLDs, which have only shared expanders, MAX 7000 EPLDs have both shared logic expanders and parallel logic expanders (see Figure 4.3.3.4).

The 16 shared logic expanders in each LAB can be viewed as a pool of uncommitted single product terms with inverting outputs that feed back into the LAB. Use of shared logic expanders enables PLA-like flexibility by allowing each shared logic expander output to be shared across all the macrocells in an LAB. Shared logic expanders can also be used to build additional register functions such as input latches.

Parallel logic expanders, on the other hand, utilize unused product terms which may be borrowed from one macrocell and assigned to an adjacent macrocell in the LAB to construct fast, complex logic. Parallel logic expanders are connected by the product-term-select-matrix in parallel with the five basic product terms in the borrowing macrocell. An additional delay of 1 to 3 nsec is incurred for this more complex logic function.

The ability to allocate additional product terms to any macrocell means that logic can be synthesized with the fewest logic resources at the fastest possible speed.

Programmable Interconnect Array

The MAX 7000 enhanced PIA is a programmable wiring path between I/O pins and LABs and from one LAB to other LABs. The PIA allows any I/O or LAB signal source to reach any destination on the device. Although it is fed by all macrocell and I/O pin feedbacks, this fast, low-skew PIA routes only those signals required to implement logic in each LAB.

The MAX 7000 PIA, shown in Figure 4.3.3.5, introduces a very short, uniform logic delay (less than 3 nsec) into the logic signal path. This has been achieved through the use of a series of two-input AND gates that feed an OR function. An EPLD bit controls one input of the AND gate and regulates the selection of the PIA signal to the LAB.

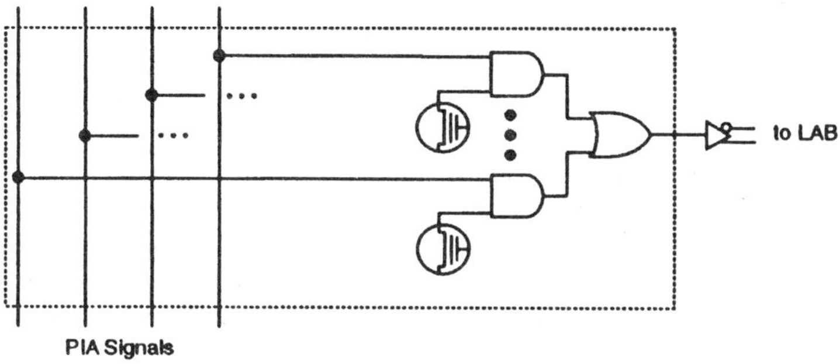

Figure 4.3.3.5 MAX 7000 Programmable Interconnect Array

I/O Control Blocks

As with other Altera architectures discussed thus far, the MAX 7000 architecture decouples I/O pins and macrocell logic resources. If an I/O pin is used as a dedicated input, the macrocell can still be used for buried logic because an independent feedback path into the logic array is provided for both the I/O pin and the macrocell.

MAX 7000 EPLDs provide two dedicated pins which can be used as global Output Enable signals. The choice of using a global Output Enable from a pin (as opposed to using a product term as was done on the MAX 5000 family) was to get maximum speed. Either OE pin can control the enable function of any output driver. The output driver can also be permanently enabled or disabled, in which case neither of the global OE signals is used. The OE input pins can also be used as general purpose logic inputs.

Routing: MAX vs. FPGAs

The ability to interconnect all points in the MAX 7000 architecture via the PIA ensures rapid, automatic design completion. Typical MAX 7000 designs can be automatically routed in minutes. Furthermore, MAX 7000 EPLDs provide a single, short uniform delay between any signal source and all signal destinations. In contrast, some programmable gate array architectures require significant manual intervention and can take hours to route. Furthermore, incremental, additive delays between various points can cause skew and glitch problems making additional iterations of the design necessary.

Programmable Speed/Power Control

The MAX 7000 family offers a programmable speed/power tradeoff that supports reduced-power operation across selected signal paths or the entire device. Other PLDs (including PALs and some members of the Altera Classic family) offer half-power or quarter-power operation across the entire device usually at the cost of reduced speed across the entire device. In the MAX 7000 family, each macrocell can be programmed by the designer for either high-speed or low-power operation. The small (less than 5 ns) speed penalty for (approximately) quarter-power operation applies only to those macrocells selected for low power. As a result, speed-critical portions of the design can run at high speed while the remainder of the design can operate at lower-power. Since only a small fraction of all gates operate at maximum frequency in most logic applications, this feature allows typical power savings of up to 50% when compared to standard PLD implementations.

4.3.4 - MPLDs: Mask-Programmed Logic Devices

General Description

Mask-Programmed Logic Devices (MPLDs) provide a masked alternative to EPLD designs. By using a generic CMOS process and removing all EPROM cells, considerable die cost savings can be achieved. MPLDs are appropriate for designs which are no longer likely to change and for which high volume is anticipated. A combination of EPLDs for prototyping and production ramp-up and MPLDs for high volume production provide both fast time-to-market and low cost and low risk. See Figure 4.3.4.1.

The EPLD-to-MPLD conversion requires no redesign effort. The MPLD is guaranteed to meet the worst-case AC and DC parameters of the original EPLD design. Test vectors are automatically generated as part of the conversion process.

EPLD/MPLD Compatibility

A MPLD is guaranteed to be pin-, function-, and timing-compatible with the original EPLD design. This guarantee ensures that the MPLD can replace the EPLD without interrupting production of the end system.

Pin compatibility guarantees that both the pin-out and DC specifications of the MPLD match those of the original EPLD design. In addition, the MPLD typically will consume less than one-tenth of the power of the equivalent EPLD, depending on the

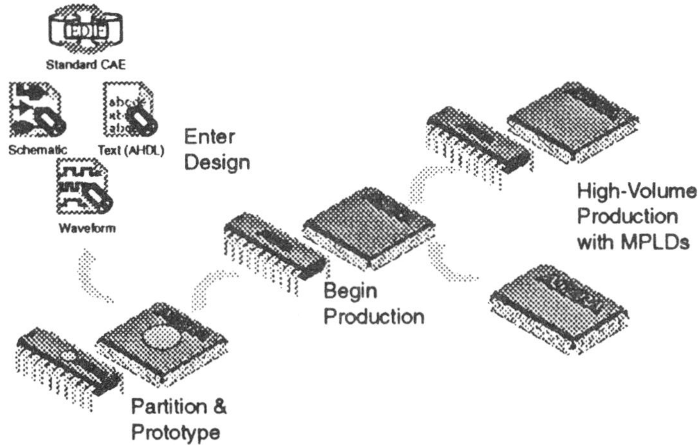

Figure 4.3.4.1 EPLD/MPLD Development Flow

design and operating conditions.

Functional compatibility of the MPLD is ensured by directly mapping the primitives within the EPLD (product terms, programmable flip-flops, etc.) to specially designed elements within the MPLD. A proprietary logic synthesis program that uses the Simulator Netlist File (SNF) generated by MAX+PLUS II software is employed in the conversion process. The SNF reflects the final synthesis, placement, routing and signal timing of the original EPLD design. The conversion process pays special attention to the wide fan-in product terms and the wide fan-out of macrocells commonly found in EPLD applications.

MPLDs are guaranteed to meet the worst-case data sheet timing parameters of the corresponding EPLD. If the design engineer performs worst-case analysis of the EPLD, the same analysis will hold for the MPLD. Therefore, the timing of the original design and the overall system is maintained when the EPLD is replaced with an MPLD.

Design for Testability

Test vector generation is one of the most time-consuming tasks required for ASIC design. A significant advantage of EPLDs over ASICs is that they are fully tested before they are shipped. The pattern programmed into the device is also verified at programming time. Application specific test vectors are not required.

MPLD designs include a partial scan-testing structure that parallels the testability available in EPLDs. The partial-scan structure allows creation of test vectors with over 95% fault coverage for all stuck-at and open faults. This high fault coverage is maintained regardless of whether synchronous or asynchronous design techniques are used. [Ahanin 1992]

The built-in design-for-testability frees the design engineer from the burden of creating a testable design and test vectors. In addition, customer-provided simulation vectors are optional and can be appended to the test vectors generated during the conversion process.

N-to-1 Conversion Option

Many applications use multiple EPLDs on a single board for both prototyping and production. In some applications it may be desirable to perform prototyping with multiple EPLDs, and then convert the design to a single-device for high-volume production (see Figure 4.3.4.1). The EPLD-to-MPLD conversion program provides this capability with the "N-to-1" conversion option, which offers the benefits of developing with EPLDs even when production constraints require a high-density single-device solution.

The N-to-1 option allows a multi-EPLD design to be converted into a single MPLD that is function- and timing-compatible with the original multi-EPLD solution. The package and pin-out is determined by the application's requirements. A wide range of package options is available.

Quick, Seamless Conversion

One of the principal objectives of the EPLD-to-MPLD conversion program is to minimize the design engineer's involvement in the conversion. With an automated conversion process, the engineer is free to begin developing the next-generation project.

The MPLD design flow chart (see Figure 4.3.4.2) shows how an EPLD is converted to an MPLD. The design engineer submits a "design packet" consisting of design files and design checklists from which a quotation is generated. Next, an engineer reviews the design and submits a Final Design Sign-off Form for customer approval. This form describes the specifications of the MPLD in detail. Following customer approval, the design conversion takes place.

The design conversion includes netlist translation, logic synthesis, testability insertion, Automatic Test Vector Generation (ATVG), fault grading, timing analysis, place-and-route, post-route timing analysis, design validation, and the manufacture of the prototypes. The entire conversion, from final design sign-off to prototype delivery, takes less than 5 weeks (6 weeks for N-to-1 conversion). Production quantities are delivered 10 to 12 weeks after the customer returns the Prototype Sign-off Form.

Summary

Two important design goals faced by engineers today are reducing time-to-market and system cost. Figure 4.3.4.3 illustrates that a combination of EPLDs and MPLDs can provide a solution that achieves these goals.

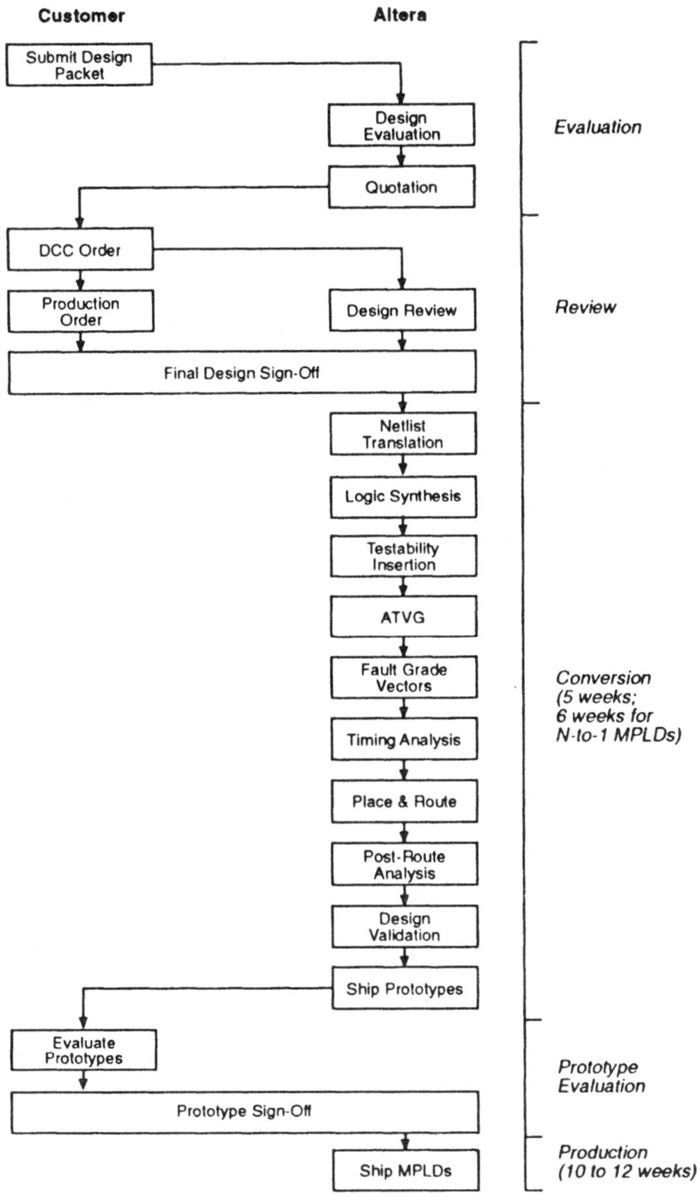

Figure 4.3.4.2 MPLD Design Flow Chart

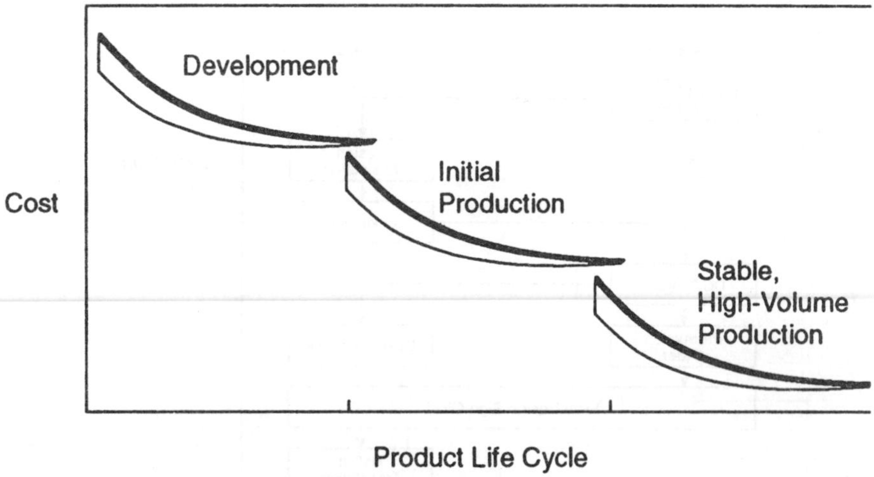

Figure 4.3.4.3 EPLD/MPLD Cost Versus Volume

4.4. Software

In previous sections the architectures of three EPLD product families were discussed in some detail. However, the philosophy of most PLD suppliers is that a logic designer need not understand the inner complexities of EPLD architectures. The user may work with familiar design entry tools (e.g., TTL functions or a high-level design language), and software should automatically translate design intent into the format required to fit the chosen EPLD architecture.

For optimum results, software products are developed together with and impose a significant influence on the EPLD architectures. Software tools that support familiar design entry methods and rapid design completion are the desired result. In this environment a user can take a logic circuit from design entry to device programming in a matter of hours (see Fig. 4.4.1). Design processing is typically completed in minutes, allowing many design iterations to be completed in a single day.

Typically, software is available for X86 PCs and Sun, HP, DEC and other workstation computers. Many design entry options are available: hierarchical schematic capture (with basic gate and complete TTL libraries), the hardware description language (AHDL, VHDL and Verilog), Boolean equation, truth table, netlist, and waveform entry. (See Figure 10.2) Design entry methods may be freely combined to create a single EPLD design. Design compilers perform minimization and logic synthesis, design fitting (analogous to automatic place-and-route), and generate programming data. Design verification via functional simulation, timing simulation, and delay prediction for speed-critical paths is also available. Hardware for programming EPLDs is offered by Altera and many PLD programmer manufacturers. Software interfaces to other design tools are provided by translators, and via industry-standard EDIF [ANSI 1988] [Datta 1991] netlists. Many third-party compilers also support EPLDs directly.

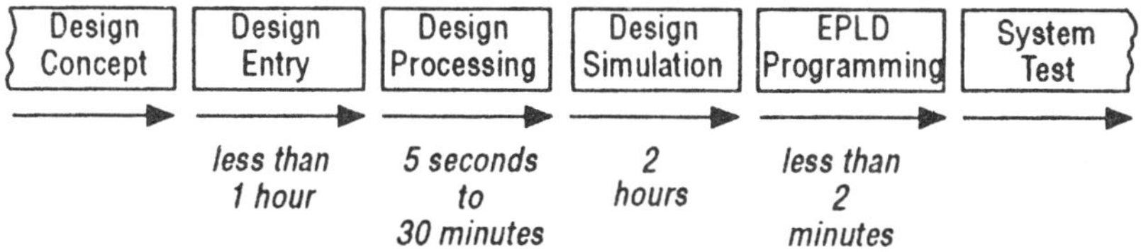

Figure 4.4.1 EPLD Design Methodology

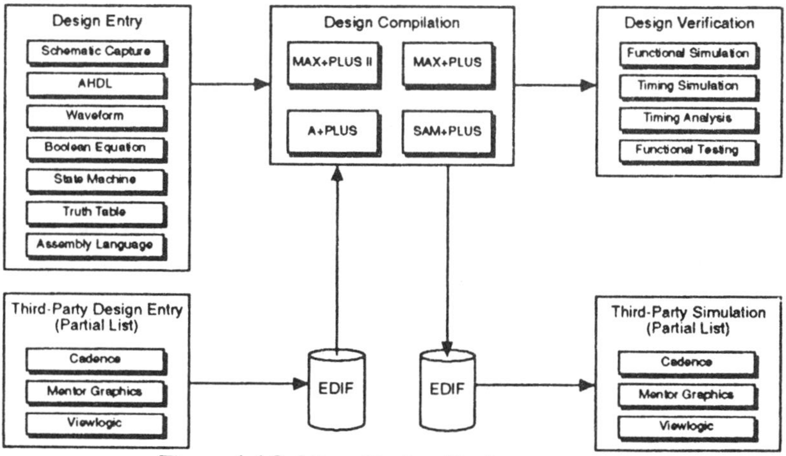

Figure 4.4.2 Altera Design Environment

MAX + PLUS II

Altera's current generation of PC-based CAE tools, MAX+PLUS II, runs in the Windows environment on PCs and in the Motif environment on workstations. The key features of MAX+PLUS II are listed below.

MAX+PLUS II Features

- Single, unified development system providing support for Altera's Classic, MAX 5000 and MAX 7000 EPLDs.
- Runs under Microsoft Windows.
  - intuitive graphical user interface
  - efficient memory management

- multi-tasking capability
- extensive printer/plotter support.

- Hierarchical graphic, text, and waveform design entry:
  - Graphic Editor for schematic-based designs
  - Text Editor for high-level textual descriptions
  - Waveform Editor for design entry and editing/viewing simulation results
- Applications run concurrently, allowing multiple editors to be active simultaneously, while simulations or compilations run in the background.
- Automatic error location is provided in the Graphic, Text, and Waveform Editors.
- Partitioning automatically divides large designs into multiple EPLDs.
- VHDL, Verilog and AHDL support.
- Logic synthesis and minimization.
- Functional simulation for detailed functional debugging.
- Interactive timing simulator supports multi-EPLD designs and probes for viewing internal nodes.
- Bidirectional Electronic Design Interchange Format (EDIF 2 0 0) netlist interface.
- On-line, context-sensitive help.

General Description

MAX+PLUS II includes design entry, compilation, multi-chip partitioning, timing simulation, and device programming support. Figure 4.4.3 shows a block diagram of MAX+PLUS II.

MAX+PLUS II supports three hierarchical design entry mechanisms: (1) schematic designs are entered with the Graphic Editor; (2) text descriptions using AHDL or EDIF 2 0 0 netlists are entered with the Text Editor; and (3) waveforms are entered with the Waveform Editor. All editors can be used concurrently, with multiple files open at a time. A library of over 300 7400-series TTL and custom macrofunctions for both schematic and text designs is included.

The Compiler synthesizes and optimizes designs using advanced logic synthesis and minimization techniques together with heuristic fitting rules to efficiently place and route the design within an EPLD. A programming file is created by the Compiler to program the EPLDs.

The multi-device partitioning automatically splits large designs (beyond the capacity

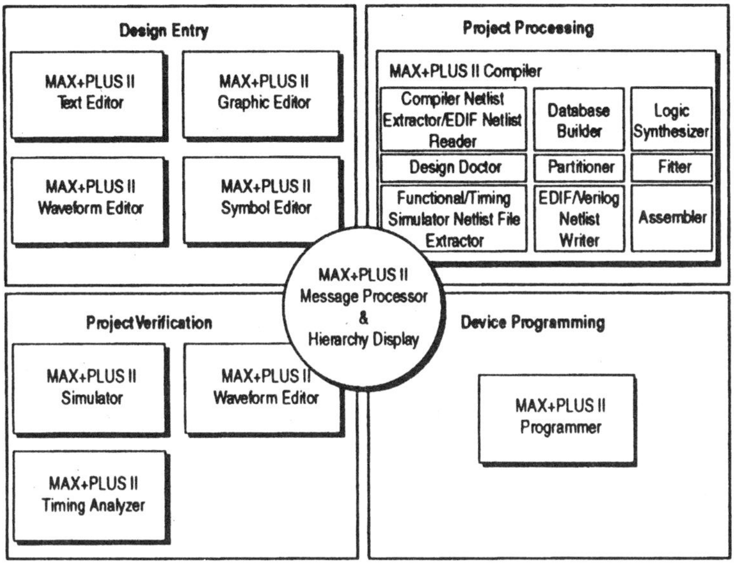

Figure 4.4.3 MAX+PLUS II Block Diagram

of a single EPLD) into multiple EPLDs, allowing the user to create large system-level designs.

The Simulator performs event-driven timing simulation. It supports multi-EPLD simulations and interactively displays timing results in the Waveform Editor. With the Waveform Editor, the user can enter, modify, and group input vectors; view simulation errors; and compare simulation runs.

The Compiler reports any design errors to the Message Processor, which automatically highlights the source of an error in the Graphic, Text, or Waveform Editor. MAX+PLUS II is fully integrated with the Windows Clipboard. The designer uses the Clipboard to quickly copy design information from one editor to another, while extensive on-line help provides instant information on all aspects of the system. The Hierarchy Display lets the designer move between hierarchical design files by simply selecting an icon.

Design Entry

Design entry files - graphic, text, and waveform - can be mixed freely. In addition to VHDL, Verilog and AHDL Text Design Files, MAX+PLUS II also accepts EDIF 2 0 0 netlists produced by CAE tools from vendors such as Cadence, Data I/O, Mentor Graphics, OrCAD, Synopsys and Viewlogic.

The Graphic, Symbol, Text, and Waveform Editors can open windows on several files at the same time. For example, a Waveform Editor window can display simulation results, while the Text Editor shows an AHDL description. At the same time, the user can open two windows of the Graphic Editor that display different levels of a design's hierarchy, or even show different areas of the same design file. If one design is displayed in two windows of an editor, any edits made in one window are automatically reflected in the other.

Graphic & Symbol Editors

The Graphic Editor (Figure 4.4.4) provides a convenient tool for schematic design entry. Designers can enter probes into the schematic to trace a specific signal (e.g., flip-flops, logic outputs) during simulation. Tag-and-drag editing can be used to quickly move individual symbols, groups of items, or entire areas. During a move, a net can be broken or connections can be preserved with orthogonal rubberbanding. Other Graphic Editor features include the ability to group nodes into buses, locate the source and destination of nets, and make quick net name changes with the search-and-replace feature.

The Compiler automatically generates a symbol that represents a design file, which can then be used in a higher-level schematic. The designer can use the Symbol Editor to modify input and output pin placement or to customize the appearance of an automatically created symbol.

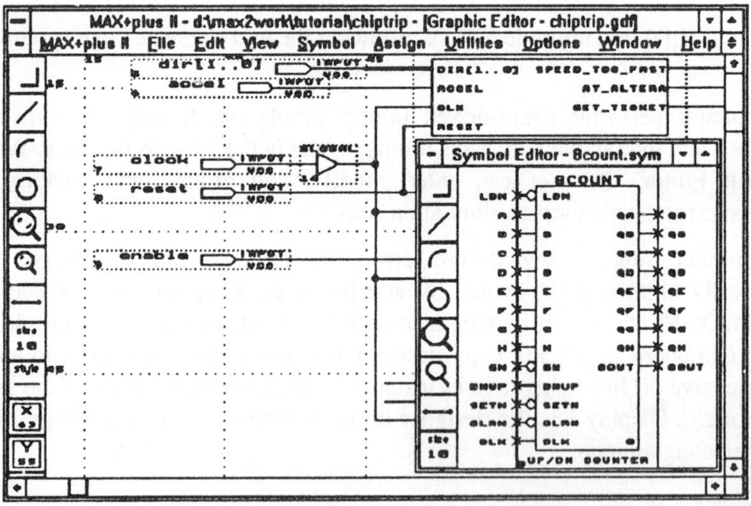

Figure 4.4.4 MAX+PLUS II Graphic Editor

Text Editor & AHDL

The Text Editor lets the user view and edit any ASCII text file in the MAX+PLUS II environment, including AHDL Text Design Files, Vector Files, Report Files, and

EDIF netlists. A Text Editor window that contains a Text Design File is shown in Figure 4.4.5. The AHDL language syntax supports arithmetic and relational operations such as addition, subtraction, equality, and magnitude comparisons. Standard Boolean functions, e.g., AND, OR, NAND, NOR, XOR, and XNOR, are also included. Since AHDL supports groups, operations can be performed on a byte- or word-wide basis as well as on single variables. AHDL also allows the designer to specify the location of nodes within Altera EPLDs. Together, these features make it easy to implement complex designs in a concise, high-level description.

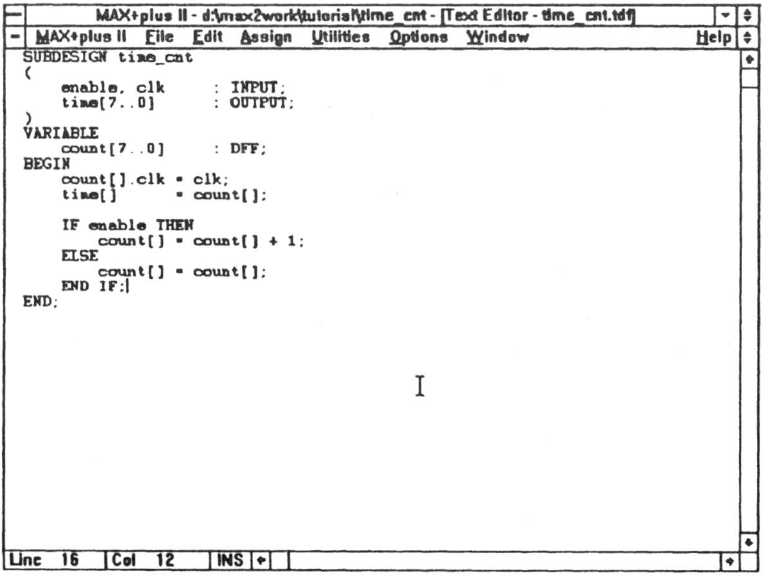

Figure 4.4.5 MAX+PLUS II AHDL Environment

Waveform Editor

The Waveform Editor (Figure 4.4.6) is used to create and edit waveform designs. In addition, the Waveform Editor functions as a logic analyzer that allows the designer to view and edit simulation results.

Designs that generate timing signals are best described with waveforms. The Compiler's waveform synthesis algorithms automatically generate logic from user-defined input and output waveforms.

Registered and combinational logic as well as state machines can be described with waveforms. The Compiler determines the optimal number of state bits and state variable assignments.

The Waveform Editor includes features to define and modify Waveform Design Files and input vectors for simulation. The designer can:

- copy, cut, paste, repeat, and stretch waveforms;
- add or delete internal nodes, flip-flops, and state machines;
- combine waveforms into binary, octal, decimal, or hexadecimal buses;
- compare the differences between two simulations by simulating a design, editing the input vectors, and then simulating the design again. The Waveform Editor superimposes the output waveforms for easy comparison.

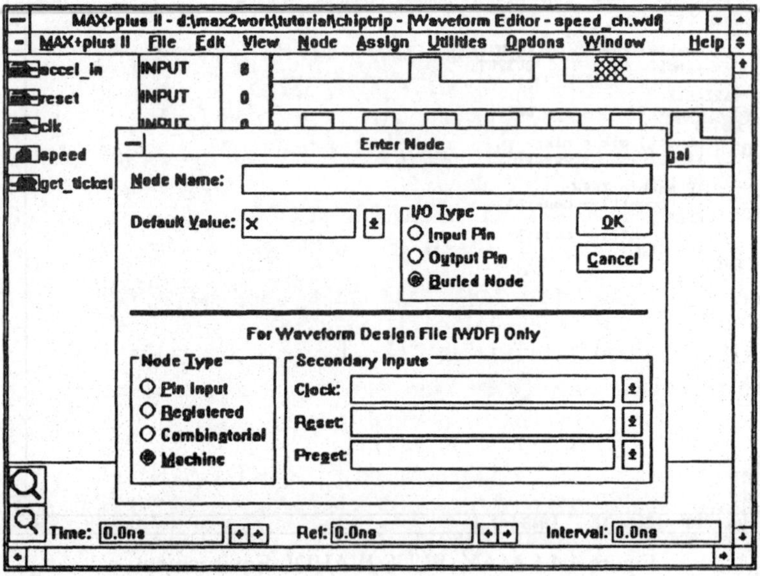

Figure 4.4.6 MAX+PLUS II Waveform Editor

Hierarchy Display

The Hierarchy Display shows the current project hierarchy - including all lower-level design files - which can be a mixture of graphic, text, and waveform files. The user selects one or more files, and MAX+PLUS II then opens the appropriate editor to display the design. This "context-sensitive editing" feature makes it easy to move around the project hierarchy.

Clipboard

The Windows 3.0 Clipboard is a temporary storage location that allows users to pass design information between editors. Text from a Text Editor file can be copied into the Graphic, Symbol, and Waveform Editors. Schematics can be copied between

Graphic Editor files, and waveforms can be pasted from one Waveform Editor file to another. Information can also be pasted into other Windows 3.0 applications.

Macrofunction Library

The MacroFunction Library contains over 300 7400-series TTL, bus, and EPLD-optimized functions. All have been optimized for speed and device utilization, and all perform true TTL emulation. Table 1 lists some of the macrofunctions currently available.

Table 1. Partial List of MAX+PLUS II Macrofunctions

Type	**Macrofunctions**
Adder	8FADD, 7480, 7482, 7483, 74183, 74283, 74385
ALU	74181, 74182, 74381, 74382
AND-OR Gate	7452
Comparator	8MCOMP, 7485, 74518, 74684, 74686, 74688
Code Converter	74184, 74185
Counter	4COUNT, 8COUNT, 16CUDSLR, GRAY4, UNICNT, 7493, 74160, 74161, 74162, 74163, 74190, 74191, 74192, 74193, 74393...
Decoder	7442, 7443, 7444, 7445, 7446, 7447, 7448, 7449, 74138, 74139, 74154, 74155, 74156...
Encoder	74147, 74148
Frequency Divider	FREQDIV, 7456, 7457
Latch	INPLTCH, NANDLTCH, NORLTCH, 7475, 7477, 74116, 74259, 74279, 74373...
Multiplier	MULT2, MULT4, MULT24, 74261...
Multiplexer	21MUX, 74151, 74153, 74157, 74158, 74298...
Parity Generator	74180, 74280
Register	7470, 7471, 7472, 7473, 7474, 7476, 7478, 74173, 74174, 74175, 74178, 74273, 74374...
Shift Register	BARRELST, 7491, 7494 ,7496, 7499, 74164, 74165, 74166, 74179, 74194, 74198...
SSI Gate	CBUF, INHB, 7400, 7402, 7404, 7408, 7410, 7411, 7420, 7421

EDIF Support

MAX+PLUS II has a built-in bidirectional EDIF 2 0 0 netlist interface, providing a convenient bridge to popular CAE schematic capture, synthesis and simulation tools. Any CAE software package that produces EDIF 2 0 0 netlists can export designs to MAX+PLUS II with Library Mapping Files (.LMF) that convert vendor CAE functions to equivalent Altera primitives and macrofunctions. Altera provides a number of ready-made LMFs for popular software packages from vendors such as Mentor Graphics, Cadence, and Viewlogic, but users can also create their own LMFs to map any CAE software library. MAX+PLUS II then automatically generates a symbol from a translated EDIF file, so that the file can be directly incorporated into a schematic or AHDL design.

EDIF netlists can also be exported from MAX+PLUS II to third-party CAE tools, allowing the user to simulate EPLD designs on the workstation of choice. Output Mapping Files can convert Altera primitives and macrofunctions to equivalent workstation functions.

Design Processing

The MAX+PLUS II Compiler processes designs for all Altera general-purpose EPLDs, including the Classic, MAX 5000, MAX 7000, and STG EPLDs (see Figure 4.4.7).

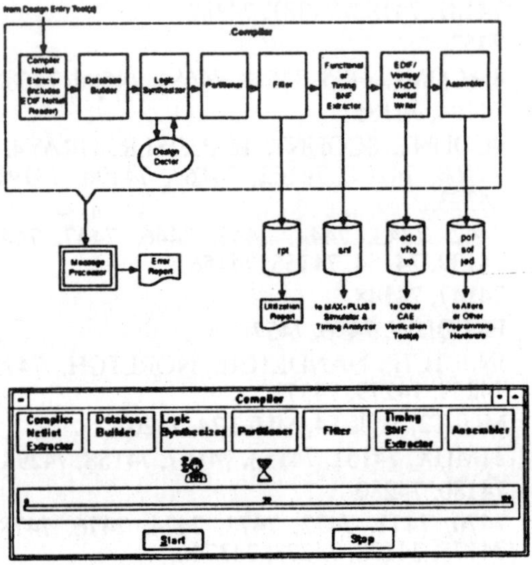

Figure 4.4.7 MAX+PLUS II Compiler

Compiler options simplify design processing and analysis. The user can specify the degree of detail of the Report File that shows how an EPLD has been utilized. The user can also specify the target EPLD family for the design, and whether or not the Compiler should extract a Simulator Netlist File for simulation within MAX+PLUS II and an EDIF Output File for third-party simulators.

The first module of the compiler, the Compiler Netlist Extractor, extracts the netlist used to define the design. This module also contains a built-in EDIF Netlist Reader. The Compiler Netlist Extractor checks design rules for any errors. If errors are found, they are passed to the Message Processor, which can then locate them in the appropriate editor. A successfully extracted design is built into a database and passed to the Logic Synthesizer.

The Logic Synthesizer module invokes algorithms that translate and optimize the user-defined logic to make the most efficient use of the resources of the target architecture. The Logic Synthesizer uses expert synthesis rules based on the target architecture (Classic, MAX 5000, MAX 7000, or STG) to factor and map logic within the chosen EPLD structure.

For simple PLD architectures, the conventional logic synthesis algorithm has been to expand the designer's logic to a sum-of-products form and then minimize the sum of products to arrive at an expression containing a minimum of product terms. If the final expression is within the product term limit for the target device, then the algorithm is successful. If not, the user must change the design and try again.

MAX-family EPLDs incorporate XOR gates and sharable, multi-level expander product terms that can be allocated wherever additional combinational logic is needed. Figure 4.3.2.3 shows a diagram of the architecture of a MAX EPLD.

The Logic Synthesizer simultaneously applies several techniques to implement design logic in the targeted device. It keeps track of the method(s) that use the fewest resources (i.e., expander product terms), and keeps only the results of the best method. The synthesis algorithm uses all combinations of techniques that are applicable in a given situation. Techniques that take advantage of the XOR gate are applied in combination with those that use the expander product terms to arrive at the best synthesis.

Eight of these logic synthesis techniques are summarized below.

1. Expansion to and Minimization of the Sum of Products

The design logic is expanded to a sum-of-products form, after which the minimization algorithm calculates the sum-of-product form with the minimum number of product terms.

2. Fitting Directly

If the logic can be expressed in a sum-of-products form without using more product terms than are available in the device, then the logic is placed directly on the product terms. Therefore, in MAX EPLDs, if the expression has three or fewer product terms, it is placed directly on the three product-term inputs of the XOR gate. (See Fig. 4.3.2.3.)

3. DeMorgan's Inversion

The Logic Synthesizer also tries both the original form and the complemented form of the design logic, in an attempt to use fewer product terms (the second input of the XOR gate is used to reinvert the logic). DeMorgan's inversion and techniques #1 and #2 above are essentially the only logic synthesis methods used in the previous generation of programmable logic compilers.

4. Fitting on Expanders

If an expression has too many product terms to fit directly, each of the product terms can be placed on an expander product term that is essentially a NAND gate. All of these expander product terms are then ANDed together on one of the product terms of the XOR gate, while the other input is used to invert the final result. (See Figures 4.3.2.3 and 4.3.2.4.) This is a straightforward NAND-NAND implementation that can always be used, no matter how many product terms there are in the sum-of-products expansion of the design logic.

5. Using XOR Gates in the Design

Instead of immediately expanding the design logic to a sum-of-products form, the Logic Synthesizer first checks whether the design uses an XOR gate as the first gate of the network of combinatorial gates. It then tries all possible ways to fit the inputs of the design's XOR gate onto the inputs of an XOR gate in a macrocell. Thus if P and Q are the inputs to the User's XOR gate, the Logic Synthesizer will try all four valid ways of placing P, Q, not P, and not Q on the "3 and 1" product term inputs of the macrocell's XOR gate (i.e., try XOR (Q, P), XOR (P, Q), XOR (not Q, not P) and XOR (not P, not Q)).

6. Factoring

The Logic Synthesizer also tries to factor a sum-of-products expression into a multi-level sum of products of sums which can be implemented on the expander product terms using NAND-AND-OR logic. The algorithm finds subsets of product terms that differ by only a single factor. These subsets are then factored into a sum of single factors ANDed with the remainder of the products. This sum is then factored out and treated as a single factor for the rest of the analysis, which continues to search for additional factorable subsets.

For example, the following sum-of-products form would naively (by technique #4) require six product terms. It can be transformed as shown below to use only two expander product terms (* is the AND operator, + is the OR operator, $ is the XOR operator, and / is the prefix inversion operator):

```
z   = a * b * c * p * x
    + a * b * c * p * y
    + a * b * c * q * x
    + a * b * c * q * y
    + a * b * c * r * x
    + a * b * c * r * y;
```

after factoring once:

```
z   = a * b * c * (x + y) * p
    + a * b * c * (x + y) * q
    + a * b * c * (x + y) * r;
```

after factoring again:

```
z    = a * b * c * (x + y) * (p + q + r);
```

and after converting to NAND for the expander product terms:

```
z    = a * b * c
       * /( /x * /y)
       * /( /p * /q * /r );
```

Thus the final expression uses only one product term and two expander product terms (the expressions in parentheses are implemented on expander product terms).

7. Finding Co-sets

This method begins with an algorithm that divides a sum-of-products expression into two subsets of product terms, such that whenever one of the subsets is true, the other is false. If two such sets are found, they are implemented on the XOR gate (when only one input is true at any given time, the XOR will function as an OR gate). The algorithm used to find two cosets of product terms starts by considering all possible pairs of product terms. If a pair of product terms has a common factor that is used in opposite polarity, then they are considered as potential seeds of the two cosets of product terms. After a pair is found, the remaining product terms are examined individually to determine whether they fit into the coset with the first or second product term. If a product term cannot go into either set, the original seed pair is discarded and another is tried until two cosets are found or until all pairs have been tried.

A 4-to-1 multiplexer provides a good example of this technique:

```
Q    = A * X * Y
       + B * X * /Y
       + C * /X * Y
       + D * /X * /Y;
```

The multiplexer fits on the XOR gate as follows (using NO expander product terms):

```
Q    = ( A * X * Y
       + B * X * /Y
       + C * /X * Y)
       $ ( D * /X * /Y);
```

8. Using the XOR as Expansion Logic

If the expression to be synthesized is the inverse of a sum of some number (N) of product terms (where N is greater than three for a MAX macrocell), this technique can always fit the logic using N-3 expander product terms. This technique relies on the following identity:

if

Q   = /(_ A_i + _ B_j);

then

Q =((_A_i) * _/B_j) $ (_/B_j);

which becomes evident when it is shown that Q is true if and only if all A_i and all B_j are false in both expressions. (A_i and B_j represent product terms.)

For example, the following inverted expression (which naively would require five expander product terms by technique #4 above):

Q =/(A * B * C + D * E + F * G + H * I + J * K);

becomes a "3 and 1" product term expression using the XOR:

Q =((A * B * C + D * E + F * G) * /X * /Y) $ (/X * /Y);

and only two expanders:

X = H * I;
Y = J * K;

For large system-level designs, the Partitioner is invoked. The Partitioner uses a sophisticated "Min-Cut" algorithm [Kernighan 1970] [Fiduccia 1982] to separate the logic design into multiple EPLDs from the same family, relieving the designer of the time-consuming task of manually splitting a large design into smaller designs. The user can control the design's partitioning by entering specific chip assignments for flip-flops and pins in the source design files.

After partitioning, the Fitter applies heuristic rules to optimally place the synthesized design into one or more chosen EPLDs. In devices with PIA structures - i.e., larger MAX 5000 and MAX 7000 EPLDs - or with local/global bus structures such as the EP1810 EPLD, the Fitter also automatically routes signals across this interconnect to relieve the designer of tedious place-and-route tasks. The Report File issued by the Fitter shows design implementations as well as any unused resources in the EPLDs.

The Simulator Netlist Extractor generates a netlist from the compiled design if the user desires simulation or timing analysis data.

The EDIF Netlist Extractor can produce an EDIF 2 0 0 netlist that contains all post-synthesis function and delay information for the completed design, so that it can be integrated into a workstation environment [Altera 1991].

Finally, the Assembler module creates one or more Programmer Object Files (POF) and/or JEDEC Files from the compiled design. The MAX+PLUS II Programmer uses these files and standard Altera hardware to program the desired EPLDs.

Message Processor

The MAX+PLUS II Message Processor is the clearinghouse for all messages generated during compilation, simulation, timing analysis, and programming. For example, if an error occurs during compilation, the Message Processor displays a brief description of the error. The user then selects the error message, chooses the

Locate button, and the troublesome logic is highlighted in the appropriate editor. Or, if a set-up time violation occurs during simulation, the Message Processor invokes not only the Waveform Editor to highlight the portion of the simulation waveform where the violation occurred, but also the appropriate editor to show the specific flip-flop location.

Functional Simulation

Functional simulation (Figure 4.4.8) allows the user to test the logical operation of a design before compilation is completed. The designer can quickly identify and correct logical errors in a design without first having to synthesize, partition, and fit the logic into an EPLD. Functional simulation is performed in the MAX+PLUS II Simulator. The Waveform Editor displays the results of functional simulation and provides easy access to all nodes in a design, including combinational functions.

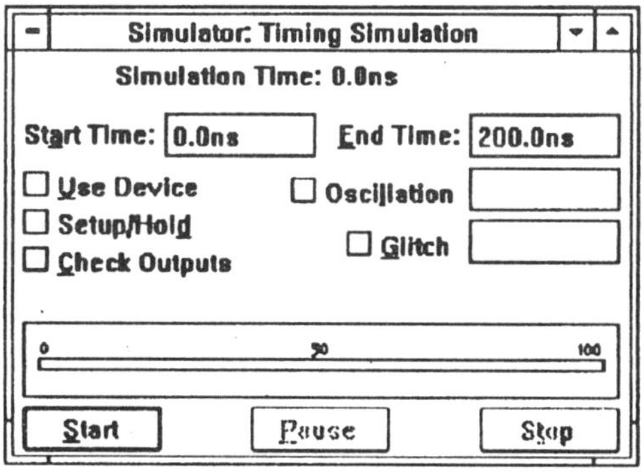

Figure 4.4.8 The MAX+PLUS II Simulator

Timing Simulation

The designer either defines input stimuli with a straightforward vector input language, or draws waveforms directly with the Waveform Editor. Simulation results can be viewed in the Waveform Editor, or printed out as table and waveform files.

The Simulator uses the Simulator Netlist File extracted from a compiled design to perform timing simulation with 0.1-ns resolution. The user can specify commands either interactively or in a batch file to perform a variety of tasks, such as halting the simulation when user-defined conditions are met or forcing flip-flops high or low.

If flip-flop setup or hold times have been violated, the Simulator warns the user, sending the information about where and when the problem occurred to the Message Processor. Also, if the user-defined minimum pulse width and period of oscillation are violated during simulation, the Message Processor locates the offending node in the original design file, and displays the time at which the problem took place in the Waveform Editor.

Differences between two simulations are viewed in the Waveform Editor, where the simulation results can be superimposed for easy comparison.

Timing Analysis

MAX+PLUS II software includes analysis tools for analyzing the timing of a completed design. The user simply tags start and end points in the Graphic, Text, or Waveform Editor to enable the Timing Analyzer to determine the shortest and longest propagation delays. The Timing Analyzer also determines setup and hold requirements at device pins, as well as maximum clock frequency. Critical paths identified by the Timing Analyzer can be highlighted in the editors.

Device Programming

All hardware and software necessary for programming, verifying and functionally testing EPLDs is available from Altera. The programming hardware includes an add-on card (for IBM PC-AT, PS/2, or compatibles) that drives the Master Programming Unit (MPU). The MPU supports functional testing, so that vectors developed during simulation can be applied to the EPLD at programming time to verify the functionality of the device. The MPU also performs continuity checking to ensure adequate electrical contact between the programming adapter and the EPLD.

In addition, Data I/O and a variety of PLD programmer manufacturers provide programming support for Altera EPLDs.

4.5. The Future

In the preceding sections we have traced the evolution of the EPLD beginning with the multi-array EP1800 and continuing through to the present day EPM7000 E-squared devices. One of the common architectural treads has been the multi-array style of architecture consisting of groups of logic elements (macrocells) which we called LABs, interconnected with a programmable routing structure which we called the PIA (programmable interconnect array). The use of EPROM technology allowed high-density, richly interconnected devices to be produced. The use of EEPROM technology allows for such devices to be programmed either off-line (as has traditionally been the case) or while in the end system. Thus, one can envision systems of EEPLDs which can be modified while in the system in the same way as is possible for SRAM programmable FPGAs.

At the same time, other programming elements, such as anti-fuses and SRAM bits, can be brought to bear on essentially LAB/PIA style architectures. One example of a new architecture currently emerging is the FLEX (Flexible Logic Element MatriX) product family from Altera. This family of devices contains many of the architectural characteristics of their predecessor EPLDs but are built on a standard CMOS technology. The programming elements for these devices are SRAM bits. A brief description of the FLEX architecture follows.

4.5.1 Functional Description

The FLEX architecture (Figure 4.5.1) incorporates a matrix of logic building blocks called logic elements (LEs). Each LE contains a four-input look-up-table (LUT) that provides combinatorial logic capability and a programmable register that offers sequential logic capability. Eight LEs are grouped together to form a Logic Array Block (LAB). LABs are arranged in rows and columns across the chip. I/O elements (IOEs) are located at the ends of rows and columns. Each IOE contains a bi-directional I/O buffer and a flip-flop that can be used as either an input or output register. Signal interconnections between LABs and to-and-from device pins are provided by an interconnect structure (called "FastTrack") made up of continuous metal lines that run the entire length and width of the device. In addition, the eight logic elements that make up a LAB communicate with each other through their own local interconnect structure without affecting the routability from one LAB to another.

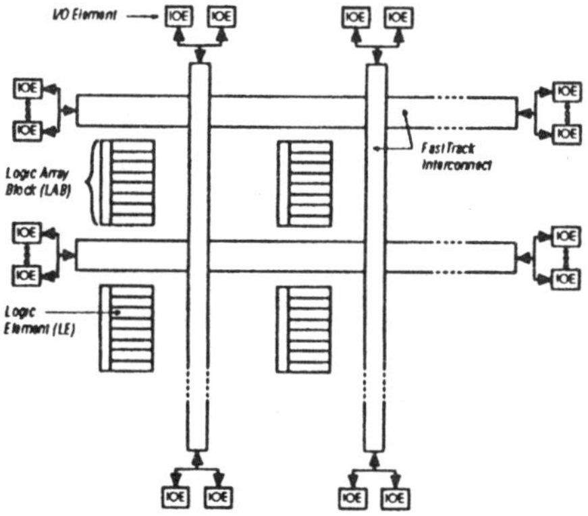

Figure 4.5.1 FLEX Device Block Diagram

Logic Elements

In addition to the LUT, each LE contains a programmable flip-flop, a carry chain, and a cascade chain. Figure 4.5.2 shows a block diagram of the LE. The programmable flip-flop in the LE can be configured for D, T, JK, or SR operation. The Clock, Clear, and Preset control signals on the flip-flop can be driven by dedicated input pins, general-purpose I/O pins, or any internal logic. For purely combinatorial functions, the flip-flop is bypassed, and the output of the LUT goes directly to the output of the LE.

Two fast data paths, the carry and cascade chains, connect adjacent LEs without using local interconnect paths. The carry chain provides a fast (less than 1 ns) carry-forward function between LEs. The carry from a lower-order bit moves forward into the higher-order bit via the carry chain, and feeds into both the LUT and the next portion of the carry chain. This feature allows the FLEX architecture to implement high-

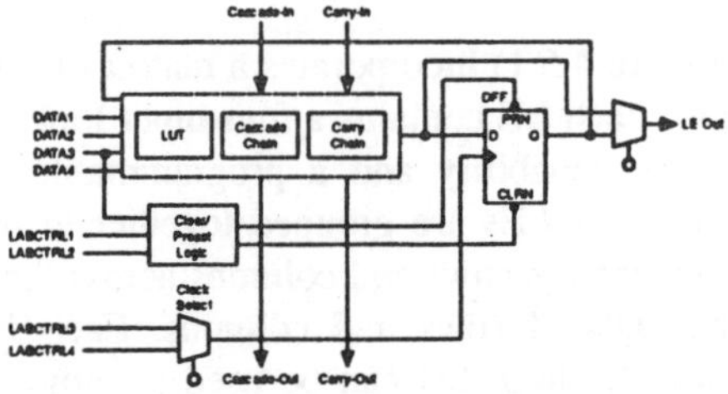

Figure 4.5.2 FLEX Logic Element (LE)

speed counters and adders of arbitrary width.

Figure 4.5.3 shows how an n-bit full adder can be implemented in n+1 LEs by using the carry chain. One portion of the LUT generates the sum of two bits using the input signals and the carry input; the sum is routed to the output of the LE. The register is typically bypassed for simple adders, but can be used for an accumulator function. Another portion of the LUT and the carry chain logic generate the carry, which is routed directly to the carry input of the next-higher-order bit. The final carry out is routed to an LE, where it can be used as a general-purpose signal.

The cascade chain provides for implementation of functions that have a very wide fan-in. The cascade chain uses a logical AND or logical OR to connect the outputs of adjacent LEs. Each additional LE provides four more inputs to the effective width of a function, with a delay of approximately 1 ns per LE. Figure 4.5.4 shows how the

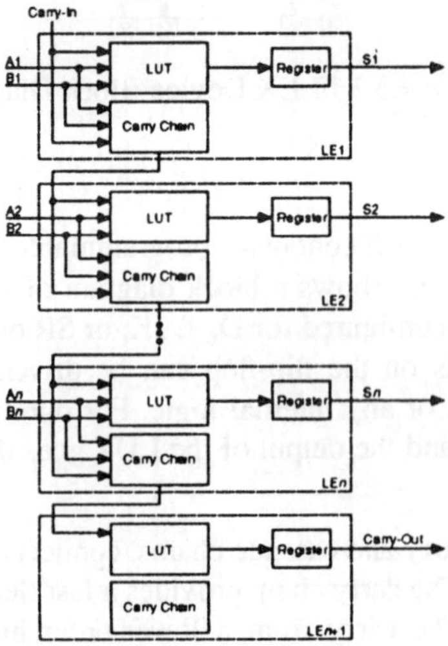

Figure 4.5.3 Carry Chain Operation

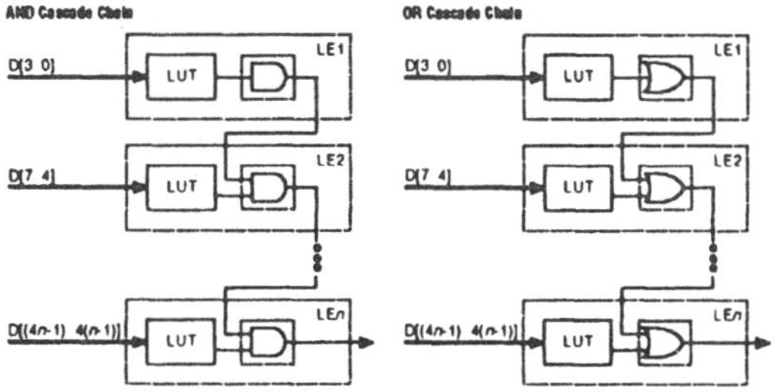

Figure 4.5.4 Cascade Chain Operation

cascade function can connect adjacent LEs to form functions with a wide fan-in. This example shows a function of 4n variables implemented with n LEs.

Logic Array Block

A Logic Array Block (LAB) consists of eight LEs, their associated carry and cascade chains, LAB control signals, and the LAB local interconnect. The LAB provides the coarse-grained structure of the FLEX architecture for efficient routing with high device utilization and high performance. Figure 4.5.5 shows a block diagram of the FLEX LAB.

Data signals enter the LAB local interconnect from either the row interconnect or the dedicated inputs. The outputs of all eight LEs are also driven back into the LAB local

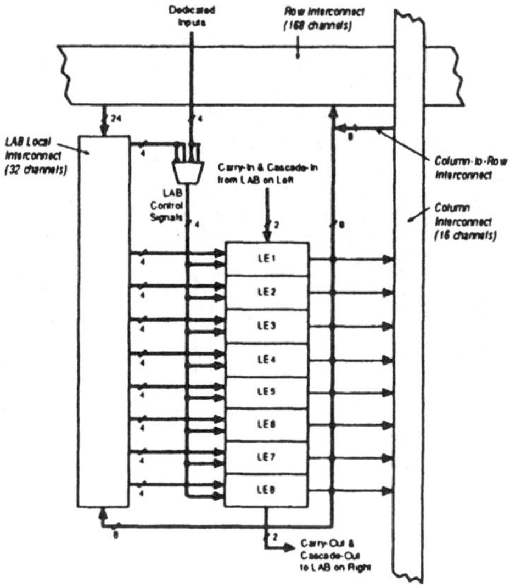

Figure 4.5.5 Logic Array Block (LAB)

interconnect via local feedback lines. Each LE in the LAB can drive signals out to the rest of the device via a row or column FastTrack interconnect path.

Each LAB provides four control signals that can be used by all eight LEs. Two of these signals can be used as Clocks, the other two as Clears. One Clear signal can also be used as a Preset. The LAB control signals can be driven directly from a dedicated input pin, an I/O pin, or any internal signal via the LAB local interconnect. The dedicated inputs are typically used for global Clock, Clear, or Preset signals because they provide synchronous control with very low skew across the device. If logic is required on a control signal, it can be generated in one or more LEs in any LAB and driven into the local interconnect of the target LAB. Programmable inversion is also available for all four LAB control signals.

FastTrack Interconnect

Connections between LEs and device I/O pins are provided by the FastTrack interconnect, a series of continuous horizontal and vertical routing paths that traverse the entire FLEX device. This device-wide routing structure provides predictable performance even in complex designs. In contrast, the typical routing structure of FPGAs consist of line segments stitched together through switch boxes, resulting in increased delays between logic resources.

Each row of LABs has a dedicated row interconnect that routes signals both into and out of the LABs in the row. The row interconnect can then drive I/O pins or feed other LABs in the device. Each column of LABs has a dedicated column interconnect that routes signals out of the LABs in the column. The column interconnect can then drive I/O pins or feed into the row interconnect to route the signals to other LABs in the device. A signal from the column interconnect, which can be either the output of an LE or an input from an I/O pin, must transfer to the row interconnect before it can enter an LAB. Figure 4.5.6 shows the interconnect of four adjacent LABs, with row, column, and local interconnects, as well as the associated cascade and carry chains.

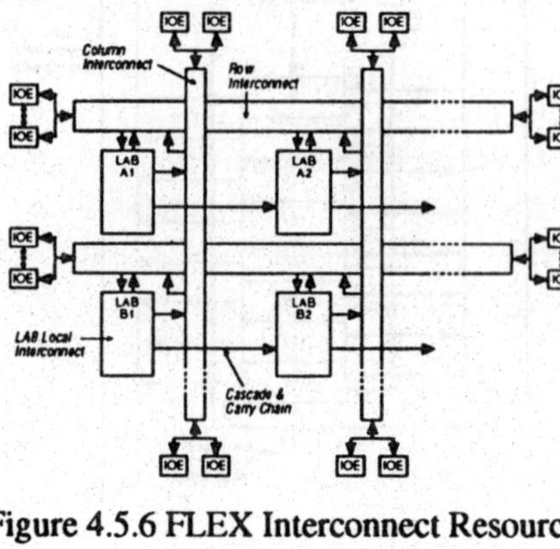

Figure 4.5.6 FLEX Interconnect Resources

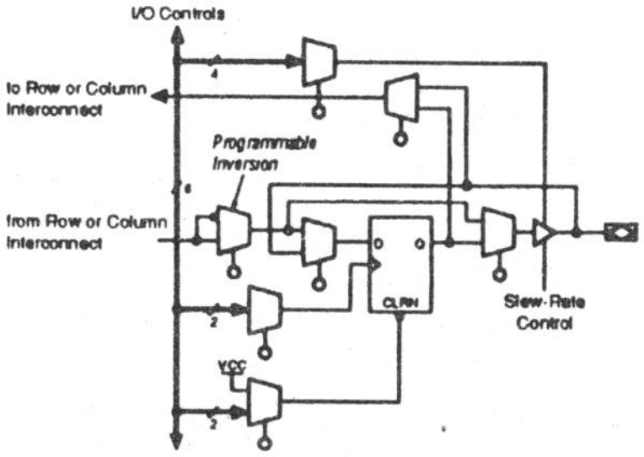

Figure 4.5.7 I/O Element (IOE)

I/O Element

Figure 4.5.7 shows the I/O element (IOE) block design. Signals enter the FLEX device from either the I/O pins that provide general-purpose input capability or the four dedicated inputs that are typically used for fast, global control signals. The IOEs are located at the ends of the row and column interconnect.

I/O pins can be used as input, output, or bidirectional pins. Each I/O pin has a register that can be used either as an input register for data that requires fast set-up times, or as an output register for data that requires fast clock-to-output performance. Each IOE has an adjustable output slew rate that can be configured for very low-noise or very high-speed performance. Designers can specify the slew rate on a pin-by-pin basis during design entry or assign a slew rate to all pins on a global basis.

The Clock, Clear, and Output Enable controls for the IOEs are provided by a network of six I/O control signals. These signals can be supplied by either the dedicated input pins or internal logic. The IOE control-signal sources are buffered to minimize skew across the device.

4.5.2 Configuration

The process of physically loading the SRAM programming data into the device is called configuration. Initialization occurs immediately after configuration. The initialization procedure resets registers, enables I/O pins, and causes the device to begin operating as a logic device. The configuration and initialization processes together are called "command mode;" normal device operation is called "user mode."

Configuration typically occurs immediately after the device is powered up but may also occur upon command. Real-time reconfiguration is performed by forcing the device into command mode with a device pin, loading different programming data, re-initializing the device, and resuming user-mode operation. The entire reconfiguration process requires less than 100 ms and can be used to dynamically reconfigure an entire system.

4.6. Design Applications

4.6.1 MAX 5000 Timing

Introduction

This section discusses internal delay paths, their relationships to AC specifications (shown in the MAX 5000 data sheets), and the calculated timing delays generated by MAX+PLUS II. Timing models for analyzing delays calculated by MAX+PLUS II, and equations that are used to calculate the delays are discussed.

EPLD Delay Parameters

Internal delays within an EPLD are described by a number of AC parameters (called microparameters) that refer to the actual internal delay within the device. Figure 4.6.1.1 shows the timing model for multiple-LAB devices. The following is a list of these microparameters and examples of how to predict timing delays with equations that use these microparameters as variables.

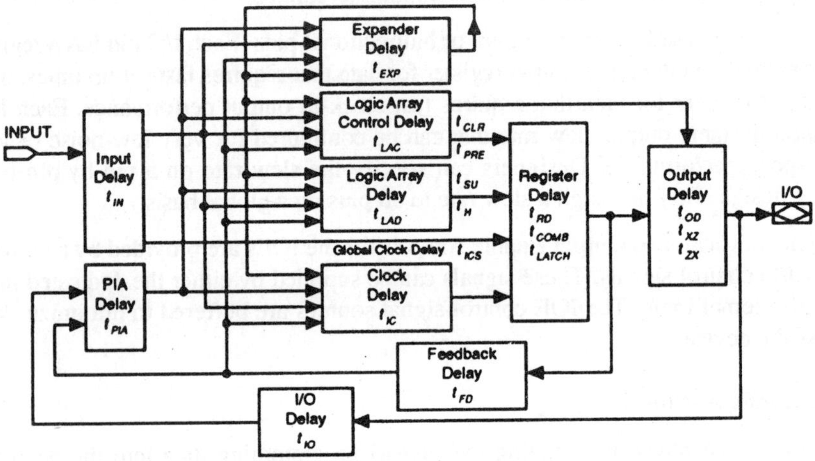

Figure 4.6.1.1 Timing Model for Multiple-LAB EPLDs

tIN Input pad and buffer delay. This delay directs the true and complement input signals from the dedicated input pin into the LAB. Within the LAB, the signals may propagate to any of four arrays: expander product-term array, logic array, logic array control, and clock array.

tIO I/O input pad and buffer delay for I/O pins used as inputs. When an I/O pin is used as an input, the tIO delay value must be added to tPIA to obtain the total delay from the I/O pin to the LAB.

tEXP Expander product-term array delay. This is the delay through the AND-NOT structure of the expander product-term array. It is added to the delay already present in the four arrays when expanders are used, or added to itself when an expander feeds another expander.

tLAC Logic array control delay. This is the delay through the AND array by Clear, Preset, and Output Enable signals, representing the time required to propagate through the AND array to the CLRN and PRN inputs to the register, and the OE signal to the tri-state buffer.

tCLR Asynchronous register clear time, which represents the time required to reset a register output to a logical low. It is the time the register CLRN input is asserted low to the time the register output stabilizes at logical low.

tPRE Asynchronous register preset time. This delay represents the amount of time required to set a register output to a logical high. It is the time the register PRN input is asserted low to the time the register output stabilizes at logical high.

tLAD Logic array delay. This is the time a signal requires to propagate through a macrocell's AND array, the three-input OR gate, and the two-input XOR gate.

tICS System clock delay. This is the delay from the dedicated clock pin to a register's clock input.

tIC Clock delay. This is the delay through a macrocell's clock product term to the register clock input.

tFD Feedback delay. This delay is the propagation time from a macrocell output to any of the LAB's arrays, or the propagation time from a macrocell output to a PIA input or other macrocells in the LAB.

tSU Setup time required for a signal to be stable at the register input before the clock's rising edge.

tH Hold time required at the register input after the register clock's rising edge to ensure that the register stores the input data.

tRD Delay from the register clock's rising edge to the time that output appears at the register output.

tCOMB Combinational buffer delay, which is used only for combinational logic. This is the delay from the time the logic array's XOR output bypasses the programmable register to the time it becomes available for the macrocell output.

tLATCH Propagation delay through the latch from latch input to output.

tOD Output pad and buffer propagation delay from the macrocell output through the tri-state output buffer to the output pin.

tXZ Delay required for high impedance to appear at the output pin after the output buffer's active-high enable control is asserted low.

tZX Delay required for the macrocell output to appear at the output pin after the output buffer's active-high enable control is asserted high.

tPIA Programmable Interconnect Array delay for multiple-LAB devices. This delay is used for designs that use the PIA for routing. The PIA delay path starts where the macrocell feedback or I/O delay path ends, and ends where it enters the LAB and reaches any of its four arrays.

Critical pin-to-pin delay calculations are shown in Figures 4.6.1.2. The calculations used to derive these values from the microparameters listed in the MAX EPLD data sheets are shown for each path. These calculations assume that a dedicated input pin is used.

If the input comes from an I/O pin, tIO is substituted for the tIN value. For multiple-LAB EPLDs that use an I/O pin as an input, tIO + tPIA is substituted for the tIN value. If an expander is used in the path at any time, the tEXP value must also be added to the total delay path.

Example: 4-Bit Counter

Figure 4.6.1.3 depicts a synchronous 4-bit counter. The counter has one clock input (CLK) and the following outputs: RCO, QD, QC, QB, and QA. In addition, it has five inherent delays associated with registered logic (clock delay, input delay, array delay, feedback delay, and output delay) as well as setup time and hold time requirements for each register.

The propagation delay from CLK input pin to the clock input of the registers is tIN+ tIC (see Figure 4.6.1.3). If an I/O pin is used for input, the propagation delay from the I/O pin to the register's clock input is tIO + tIC, or tIO + tPIA + tIC for MAX EPLDs with multiple LABs. Since the delay from register to output pin is tRD + tOD, the total clock to output delay is tIN + tIC + tRD + tOD for dedicated input to output; tIO + tIC + tRD + tOD for I/O pin to output for MAX 5000-series EPLDs with one LAB; or tIO + tPIA + tRD + tOD for I/O pin to output for MAX 5000-series EPLDs with multiple LABs.

In addition, data input to the register must meet both setup and hold time requirements. The internal setup time is the time needed for the input data to stabilize before the triggering edge of the clock appears at the register input. The external setup time is the algebraic difference between the sum of the input, logic array, and setup time, and the sum of the input and clock delay: (tIN + tLAD) - (tIN + tIC) + tSU. When expanders are used, tEXP must be added, and when I/O pins are used, tIO or (tIO + tPIA) must be added. As long as the external setup time is met at the inputs, the counter functions properly.

The maximum internal counter frequency (fCNT) is the inverse of tCNT , which is the worst-case delay for internal feedback. This frequency is the minimum internal clock period at which the counter can operate correctly. tCNT is the sum of delay

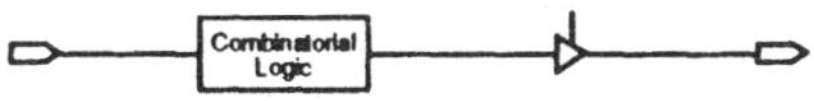

MAX 5000 $t_{PD1} = t_{IN} + t_{LAD} + t_{COMB} + t_{OD}$

(multi-LAB) $t_{PD2} = t_{IO} + t_{PIA} + t_{LAD} + t_{COMB} + t_{OD}$

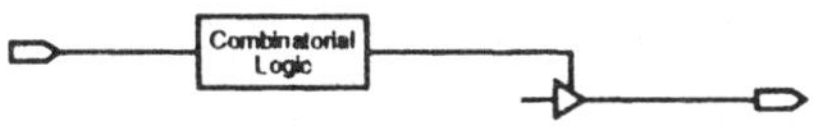

MAX 5000 $t_{PXZ}, t_{PZX} = t_{IN} + t_{LAC} + (t_{XZ} \text{ or } t_{ZX})$

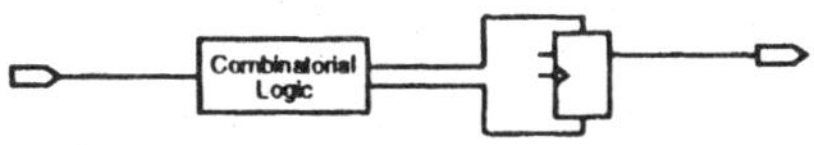

MAX 5000 $t_{PRE}, t_{CLR} = t_{IN} + t_{LAC} + (t_{PRE} \text{ or } t_{CLR}) + t_{OD}$

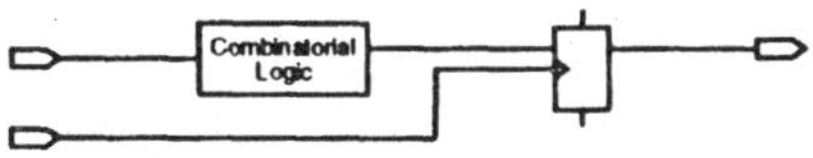

MAX 5000 $t_{SU} = (t_{IN} + t_{LAD}) - (t_{IN} + t_{ICS}) + t_{SU}$

MAX 5000 $t_{H} = (t_{IN} + t_{ICS}) - (t_{IN} + t_{LAD}) + t_{H}$

MAX 5000 $t_{CO1} = t_{IN} + t_{ICS} + t_{RD} + t_{OD}$

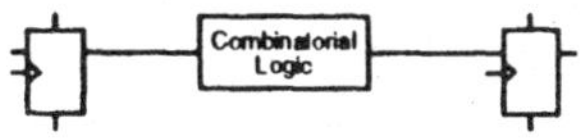

MAX 5000 $t_{CNT} = t_{RD} + t_{FD} + t_{LAD} + t_{SU}$

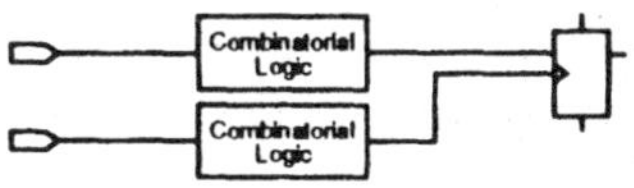

MAX 5000 $t_{ASU} = (t_{IN} + t_{LAD}) - (t_{IN} + t_{IC}) + t_{SU}$

MAX 5000 $t_{AH} = (t_{IN} + t_{IC}) - (t_{IN} + t_{LAD}) + t_{H}$

MAX 5000 $t_{ACO1} = t_{IN} + t_{IC} + t_{RD} + t_{OD}$

Figure 4.6.1.2 Critical Pin-to-Pin Delay Calculations

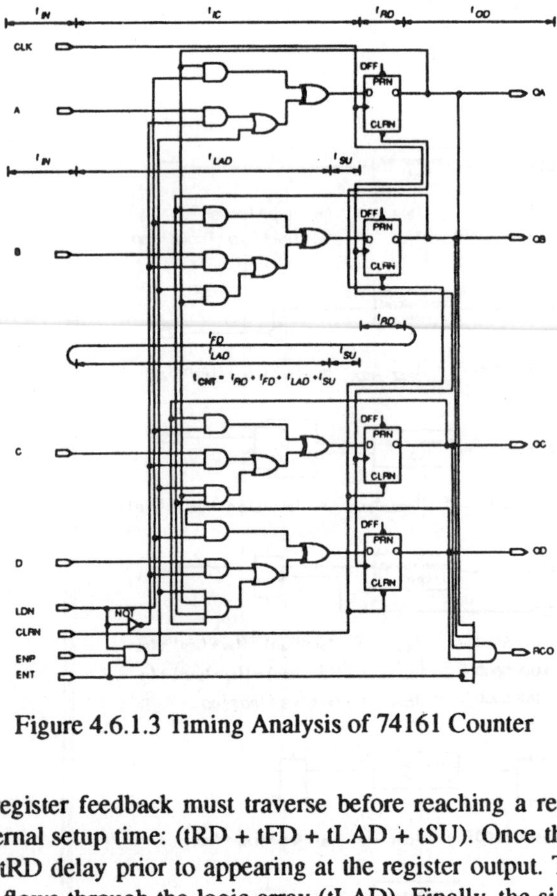

Figure 4.6.1.3 Timing Analysis of 74161 Counter

paths that the register feedback must traverse before reaching a register input and meeting the internal setup time: (tRD + tFD + tLAD + tSU). Once the clock triggers QB, data takes tRD delay prior to appearing at the register output. The signal feeds back (tFD) and flows through the logic array (tLAD). Finally, the signal reaches the register QC and meets the setup time of the register (tSU).

When expanders are used, tEXP must be added as the signal passes through the expander array before reaching the logic array. The tCNT delay represents only internal circuit delays, while a circuit that depends on external and internal signals must also account for input and I/O delays. For example, the internal path (tCNT) is 20 nsec for EPM5128-1, while the external path (tSU + tCO) is 29 nsec.

4.6.2 Using Expanders to Build Registered Logic in MAX EPLDs

Introduction

Each EPLD macrocell contains one register that can be programmed for registered functions or bypassed for combinational functions. However, some applications require more registers than are available in the macrocells. These extra registers can be built with expander product terms (expanders). This section explains how and when to use expander latches and registers, and describes timing considerations, specifically for an SR latch, a transparent D latch, and a synchronous register.

Expander product terms

Expanders are unallocated product terms with inverted outputs that feed the logic array. Each expander is fed by the same inputs that feed the macrocell: a global bus, macrocell feedbacks, other expanders, and I/O feedbacks. Since expanders feed themselves, they can be used to build latches and registers. Two expanders (EXP primitives) can be cross-coupled to generate an SR latch, three can be used to build a transparent D latch, and six can be used to build a synchronous D flip-flop with asynchronous Preset and Clear. The expander circuits described here have been built into macrofunctions to optimize performance. Each function is built with AND gates and EXP primitives to optimize fitting.

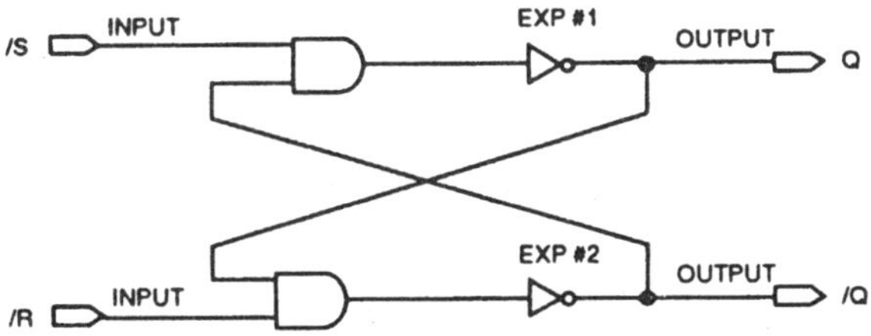

Figure 4.6.2.1 SR Latch Implemented with Expanders

Asynchronous SR Latch

Figure 4.6.2.1 shows an asynchronous SR latch implemented with two expanders. Since expanders are product terms with inverted outputs, the latch is a NAND implementation with the Set and Reset terms active low. If both inputs are simultaneously low, both outputs will become logic high until either of the inputs goes high.

The functional output of an SR latch is shown in Table 1. To implement the latch with active-high inputs, as in a NOR latch, the inputs are inverted with NOT primitives. Asynchronous SR latches are often used to debounce input-switching circuits or to detect edges in switching circuits.

Table 1

/S	/R	Q
H	H	Q0
L	H	H
H	L	L
L	L	H (1)

Note:(1) /S and /R low causes both Q and /Q to be high.

SR Latch Timing

Each expander has a timing delay defined as tEXP. The hold time for the SR latch shown in Figure 4.6.2.2 is given as tH = 2* tEXP. A low signal at the S input must remain low long enough to propagate through Expander 1 and Expander 2 to latch the input. The propagation delay from the latch input to the latch output is given as tCO = 2* tEXP. A low at the S input must travel through Expander 1 and Expander 2 before the outputs Q and /Q become valid.

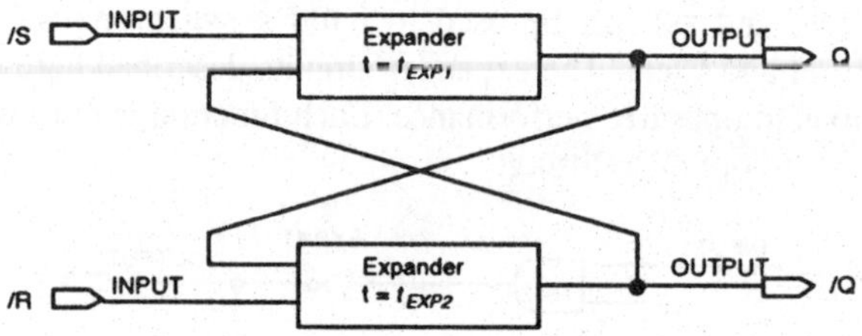

Figure 4.6.2.2 SR Latch Timing Model

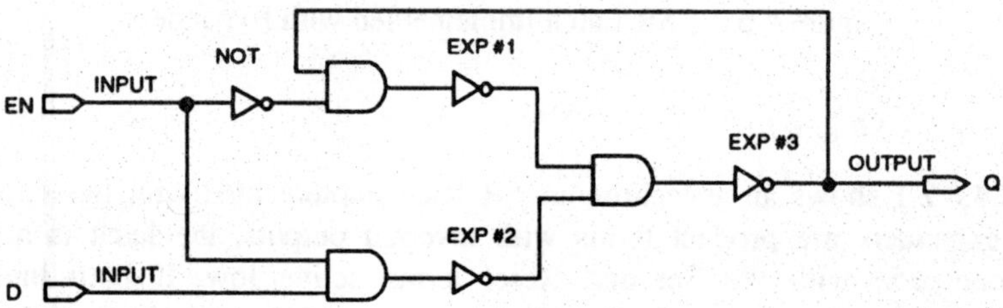

Figure 4.6.2.3 Transparent D Latch Implemented with Expanders

Transparent D Latch

The transparent asynchronous D latch with EN (Enable) is implemented with three expanders, as shown in Figure 4.6.2.3. This latch is functionally comparable to a 74LS373. The latch is transparent when EN signal is logic high. When EN goes low, the input is latched. This latch is especially applicable for latching inputs from a bus. The functional output of this latch is shown in Table 2.

Table 2

EN	D	Q
L	L	Q0
L	H	Q0
H	L	L
H	H	H

Transparent D Latch Timing

The delay paths for the transparent asynchronous D latch are shown in Figure 4.6.2.4. The tH value for this circuit is 0 ns because the paths from the D input and the EN input have delays equal to those of Expander 3, which latches the result. Both the set up time tSU and the enable to output time tCO is given as two expander delays. The setup time, tSU, requires the D input to go through Expander 2 and Expander 3 to reach Expander 1 before it can be latched. tCO is the delay through Expander 2 and Expander 3 to the output from the rising edge of EN.

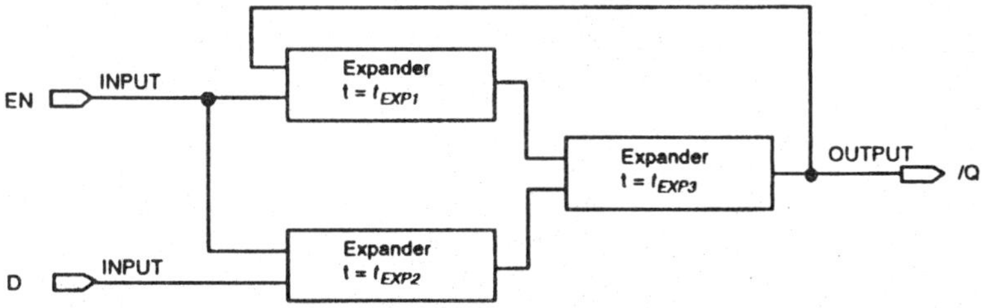

Figure 4.6.2.4 Transparent D Latch Timing Model

Synchronous D Register

The function outputs for the synchronous D register are shown in Table 3. A synchronous D register with asynchronous Preset and Clear (/P and /C) can be built with six expanders, as shown in Figure 4.6.2.5.

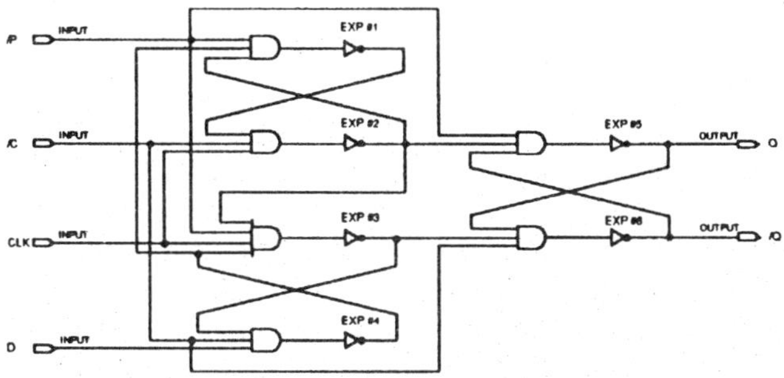

Figure 4.6.2.5 Synchronous D Register with Preset and Clear

Table 3

/P	/C	D	CLK	Q	/Q
H	H	L	U	L	H
H	H	H	U	H	L
H	H	X	H	Q0	/Q0
H	H	X	L	Q0	/Q0
L	H	X	X	H	L
H	L	X	X	L	H
L	L	X	X	H	H

The state at the D input is clocked into the latch with a rising edge at the clock input. Both the true and complement signals are available at the output. This output remains latched until the next rising edge clock or until the Preset or Clear is asserted low. The Preset and Clear can be made active high by placing NOT primitives in front of the two signals. Using expanders as registers increases the total register count in a given MAX 5000 device by 31%. For example, the EPM5192 can have up to 60 registers implemented with expanders, which gives it a total capacity of 252 registers. However, the speed of these expander registers will be slower than the macrocell registers.

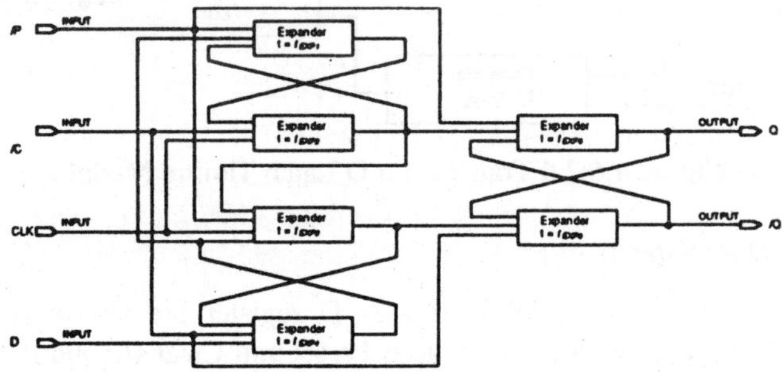

Figure 4.6.2.6 Timing Model for Synchronous D Register

Synchronous D Register Timing

The timing paths for the synchronous D register are shown in Figure 4.6.2.6. Two different timing paths exist for the D input, depending on whether the D input is high or low. The worst-case path is described here. The worst case th is one expander delay through Expander 3. The worst case for tsu is the path through Expander 4 and Expander 1 to the input of Expander 2. The worst case clock-to-output (called tco1) path is through Expander 3, Expander 6, and Expander 5 for both Q and /Q to produce valid outputs. Both tclr and tpre have a delay path through Expander 5 and Expander 6 to produce a valid output. tcnt is the minimum recirculation time for the register which is tco1 + tsu or five expander delays. The inverse of tcnt is fcnt or the maximum toggle frequency.

Fitting Expanders Into MAX Designs

Latches and registers built using expanders can provide valuable additional resources. Their use, however, must be carefully analyzed to ensure correct setup time, hold time and tco. These expander latches and registers are particularly useful for latching input signals. They should not be used when the D input is a complex function because other expanders are the only source of this logic. Instead, registers driven by complex logic should be placed in macrocells. On the other hand, if additional logic is required after the register, expander registers should be used, since the output feeds directly into the logic within the macrocells.

4.6.3 Simulating Internal Buses in General-Purpose EPLDs

Introduction

Altera's EPLDs allow internal buses to be emulated by using logic to replace tri-state functions. A series of simple multiplexers can create buses with several sets of input signals. Multiplexing also saves device resources and helps to eliminate timing and loading problems. This section describes how to use multiplexers for different bus configurations and explains the benefits of this approach.

Using multiplexing to emulate tri-state functions saves macrocells and I/O pins for applications that would otherwise require a bus external to the EPLD. Figure 4.6.3.7 shows a 4-to-1 multiplexer in a single macrocell that emulates a bus line with four sources. With conventional tri-stating techniques, the same function requires four macrocells and I/O pins, as shown in Figure 4.6.3.8. Multiplexing saves three macrocells and I/O pins if the switching functions are implemented with the product terms inside the macrocell, instead of with tri-state buffers and I/O pins external to the macrocell.

Two-Source Bus Configurations

Figure 4.6.3.1 shows the simplest bus configuration, a one-bit bus created by connecting the outputs of two tri-state buffers to a single line. The function table shows the possible states of the bus. When tri-state buffer A is enabled, the input to

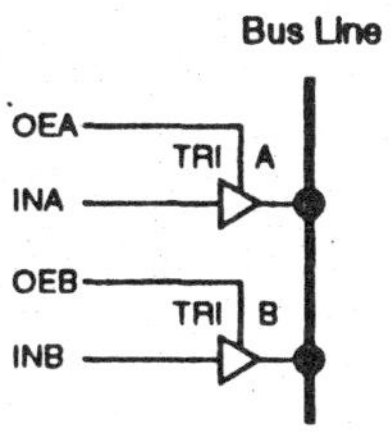

INA	OEA	INB	OEB	BUS LINE
X	0	X	0	Z or 1
X	0	0	1	0
X	0	1	1	1
0	1	X	0	0
1	1	X	0	1

Figure 4.6.3.1 One-Bit Bus

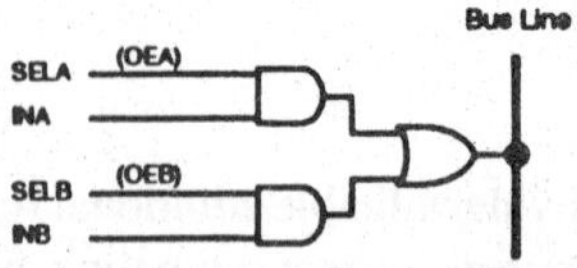

INA	SELA	INB	SELB	BUS LINE
X	0	X	0	0
X	0	0	1	0
X	0	1	1	1
0	1	X	0	0
1	1	X	0	1

Figure 4.6.3.2 AND/OR Logic Emulating Tri-State Functions

that buffer (INA) appears on the bus. When tri-state buffer B is enabled, the input to that buffer (INB) appears on the bus. If neither buffer is enabled, the bus is in a high-impedance, or floating, state. Such buses are often tied high with a pull-up resistor to prevent them from floating.

Figure 4.6.3.2 shows two AND gates and an OR gate that emulate the tri-state functions of Figure 1. Each AND gate has a data input (INA or INB), and a select input (SELA or SELB) that represents the original Output Enable control. The function table shows that the AND/OR logic exactly emulates the original tri-state functions, if one of the two outputs is always selected. If neither output is selected, the output of the AND/OR logic is low.

The select controls are mutually exclusive, since only one input is ever enabled onto a bus at any given time. Therefore, they can be encoded into a single input by making SELA the common select input, and then feeding the inverse of this signal into the previous SELB input as shown in Figure 4.6.3.3.

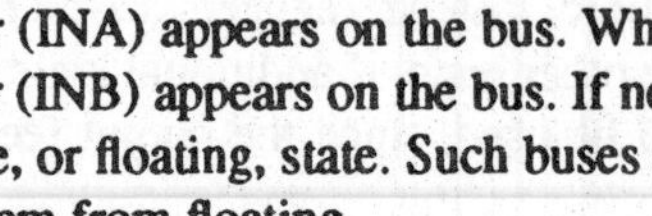

INA	INB	SEL	OUTPUT
X	0	0	0
X	1	0	1
0	X	1	0
1	X	1	1

Figure 4.6.3.3 Multiplexer Created with AND/OR Logic and Select Controls

Additional 2-to-1 multiplexers, all controlled by a common select signal, can create wider buses. One multiplexer is necessary for each bit of the bus. For example, Figure 4.6.3.4 shows eight 2-to-1 multiplexers emulating a byte-wide bus.

Buses with Three or More Sources

Larger multiplexers with multiple select inputs can emulate buses with more than two sources. Figure 4.6.3.5 shows how a 4-to-1 multiplexer can create a bus with up to four sources. Figure 4.6.3.5 also includes a truth table with the proper encoding for the select inputs. This type of multiplexer can also implement buses with two or three sources.

Additional multiplexers with shared select lines can create buses with nearly any width. For example, five 4-to-1 multiplexers can create a 5-bit-wide bus with two, three, or four sets of inputs.

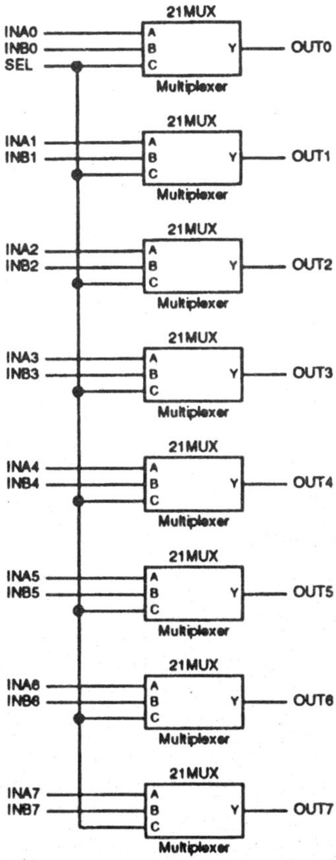

Figure 4.6.3.4 Eight 2-to-1 Multiplexers Emulating a Byte-Wide Bus

Implementing Bus Functions with Hardware Description Language

Hardware Description Languages, such as Verilog, VHDL (VHSIC Hardware Description Language), or AHDL [Altera AN22 1990] provide a quick alternative to graphic schematic entry for implementing bus functions with multiplexing. HDLs can be used to describe buses with nearly any number of inputs and of nearly any width.

For example, Figure 4.6.3.6 (next page) shows the AHDL code required to create an eight-bit bus with three sources. The data inputs are A7 to A0, B7 to B0, and C7 to C0. The two select inputs, SEL1 and SEL2, can be treated as an encoded group in AHDL. These select lines control which set of input signals is connected to the outputs through a series of simple IF-THEN statements.

By adding data inputs (e.g., D[7..0]), this file can be easily modified to create a bus with more sources. For multiplexers with more than four data inputs, one more bit

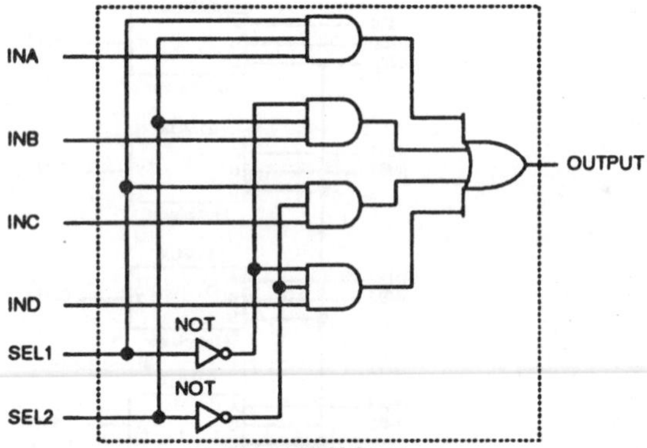

INA	INB	INC	IND	SEL2	SEL1	OUTPUT
X	X	X	0	0	0	0
X	X	X	1	0	0	1
X	X	0	X	0	1	0
X	X	1	X	0	1	1
X	0	X	X	1	0	0
X	1	X	X	1	0	1
0	X	X	X	1	1	0
1	X	X	X	1	1	1

Figure 4.6.3.5 Four-to-One Multiplexer Implementing a Bus with up to Four Sources

```
SUBDESIGN BUSMUX (
    A[7..0],
    B[7..0],
    C[7..0],
    SEL[1..0]: INPUT;
    OUT[7..0]: OUTPUT; )
BEGIN
    IF (SEL[]==0) THEN OUT[]=A[]; END IF;
    IF (SEL[]==1) THEN OUT[]=B[]; END IF;
    IF (SEL[]==2) THEN OUT[]=C[]; END IF;
END;
```

Figure 4.6.3.6 AHDL Implementation of Eight-Bit Bus with Three Sources

must also be added to the SEL group (e.g., SEL[2..0]) for each factor-of-two increase in the number of data inputs. For example, a seven-input multiplexer requires three select bits.

The width of the bus can be varied by changing the input and output group widths. For example, the declaration A[5..0] creates a 5-bit wide set of A inputs.

PROs and CONs

Emulating tri-stated buses with logic eliminates timing hazards such as bus contention, which occurs when two or more tri-state outputs are simultaneously enabled onto a single bus line. This condition (usually unintended) can cause an unpredictable logic level to propagate if multiple buffers are driving high and low at the same time. The select controls for simple AND/OR logic (shown in Figure 4.6.3.2) can both be enabled at the same time, but the result will be a known logic level. The select controls can never be enabled at the same time if they are encoded, as in the true multiplexer configurations (Figure 4.6.3.3).

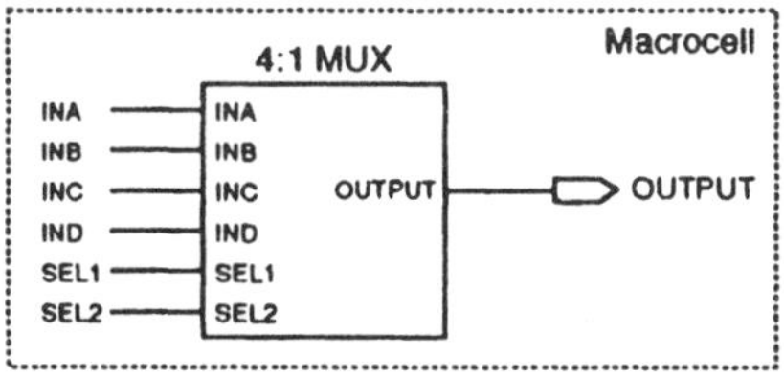

Figure 4.6.3.7 Four-to-One Multiplexer in a Single Macrocell

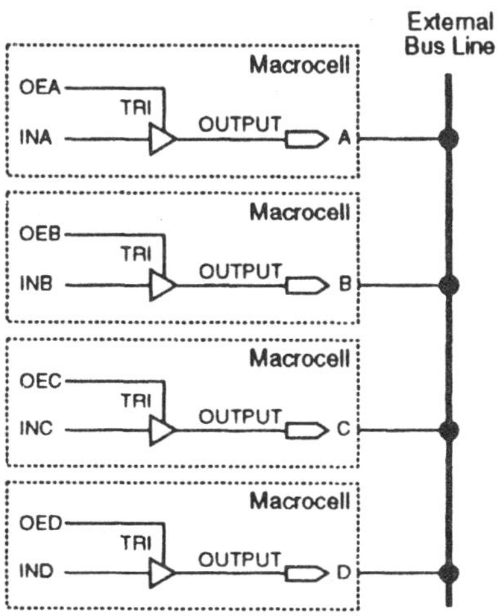

Figure 4.6.3.8 Four-to-One Multiplexer Implemented with Traditional Tri-State Logic

A potential timing hazard for a multiplexer configuration is output glitching caused by input signal skew. EPLD architecture minimizes skew difficulties and glitching is seldom a problem in actual designs. However, the designer must exercise care when driving edge-sensitive logic from multiplexer outputs. A potential disadvantage of replacing tri-state buses with multiplexers is that all signals must be routed to the single multiplexer location. If the function can be implemented in a single EPLD, this is usually not a problem. If the tri-state function is spread across several chips, then signal routing is a more serious problem.

Replacing tri-stated buses with logic reduces capacitive loading limitations. High fan-outs to traditional buses create high capacitive loads that reduce bus bandwidth. Macrocells and feedback paths in EPLDs have constant delays, regardless of the number of signals entering the macrocell. If control logic is implemented with multiplexers, internal loading is not a problem.

4.6.4 Fast Bus Controllers with the EPM5016

Today's advanced microprocessors are capable of running in systems with clock speeds greater than 33 MHz. To realize the performance potential of these microprocessors, their memory interfaces must be equally fast. High-performance memory devices, however, are expensive. Cost can be decreased with little impact on performance by using a combination of fast and slow memories. To accommodate the slower memories, wait states are added into the microprocessor bus cycle. A bus controller which contains the wait-state and bus-control logic for a 80386 system can be integrated into an Altera EPM5016 MAX EPLD. The EPM5016 has a propagation delay of 15 ns and can support a system clock rate of 66 MHz. With a specified output drive of 24 mA, it can be directly connected to buses.

Figure 4.6.4.1 shows a block diagram of an 80386 microsystem that incorporates peripheral logic, memory, and an 8259A interrupt controller. The EPM5016, shown in the center of the diagram, serves as the system bus controller. The EPM5016 decodes the 80386 status signals to control the peripheral logic, the data transceiver, interrupt controller, and other external logic. It also extends bus cycles by adding wait states to interface to slower peripherals and memory devices.

The 80386 halts processing to allow wait states to be added into the bus cycle when the signal /READY is high. (Note that the slash (/) is used to indicate an active-low signal.) The EPM5016 bus controller tracks each bus cycle operation and causes /READY to go high when wait states are needed. For example, read operations from 200-ns EPROM memory in 33-MHz systems require 14 wait states.

The EPM5016 also decodes the bus control signals IORD (I/O read), IOWR (I/O write), and INTA (interrupt acknowledge). The 24-mA output drivers on the EPM5016 eliminate the need to buffer these bus signals externally.

The 16 data signals originating from the 80386 are isolated from the system data bus

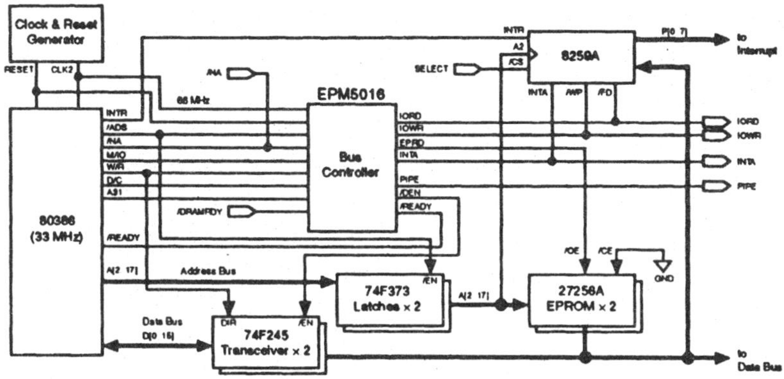

Figure 4.6.4.1 80386 Subsystem Block Diagram

with two 74245 8-bit transceivers. The tri-state control on the transceivers is provided by the signal /DEN from the EPM5016. The direction signal is controlled directly by the Read-Write signal (W/R).

Two 74373s (8-bit latches) are used to latch the 80386 address signals at the beginning of the bus cycle to maintain a valid address throughout the cycle. The latches are controlled by the 80386 signal /ADS. The high performance of the EPM5016 easily supports the 33-MHz bus cycles. In fact, a 66-MHz (2 x 33-MHz) clock is used to clock the design for two reasons. First, /ADS can be connected directly to the address latches since the EPM5016 control signals for the peripheral logic are active before the end of the first bus cycle. Second, the wait-state generator offers finer granularity with 15-ns cycles than with 30-ns cycles.

The design for the bus controller circuit requires 9 inputs: all 8 dedicated inputs of the EPM5016 and 1 I/O pin. Five of the 9 inputs to the EPM5016 are signals from the 80386 microprocessor. The functions of these 5 signals are given in Table 1. The other inputs are described subsequently.

Table 1. EPM5016 Input Functions

Input	Function
M/IO	Memory or I/O
W/R	Read or Write status
D/C	Data or Control status
A31	Address bit 31 for memory mapping of the EPROM
/ADS	Address data strobe indicating the beginning of the bus cycle

Systems that have functions set up for pipelining require external logic to generate the signal /NA (next address). /NA feeds both the 80386 and the EPM5016 bus controller. When /NA is activated, the 80386 places the next address on the address bus so that it may be latched. Applications that require wait states receive minimal benefit from pipelining. In such cases, the /NA signal is used simply to disable the EPM5016.

/DRAMRDY is an externally generated signal that, when high, halts the 80386 by causing the /READY signal to be high. When /DRAMRDY goes low, the 80386 continues processing.

The 66-MHz clock (CLK2) feeds a toggle flip-flop in the EPM5016 to generate a divide-by-two signal (CLK) that matches the 33-MHz system clock signal. CLK tracks the microprocessor clock phase.

The last input signal, RESET, is connected directly to the reset signal of the microprocessor. RESET feeds the preset or reset of each register to set the EPM5016 to the correct start-up state.

Figure 4.6.4.2 shows the flow diagram for the bus controller. First, the 80386 causes /ADS to go low, indicating that a bus cycle has begun. /ADS then causes BUS_TRCK to activate the signal BUS_ACTIVE, which indicates that the processor is in an active bus cycle. BUS_ACTIVE feeds the functions DECODE and TRAN_CTL.TRAN_CTL then scans the 80386 control signals and enables the data transceiver buffers when they are required. DECODE also scans the 80386 control signals and decodes them into specific control signals (IORD, IOWR, EPRD, and INTA) that drive the peripheral logic and the block function, WAIT_CNT.WAIT_CNT is a loadable 4-bit counter that counts the required number of wait states for bus cycles. When WAIT_CNT finishes the wait-state count, it causes the TIME_DELAY to go low, enabling DECODE to release the /READY signal and finish the bus cycle.

4.6.5 Micro Channel Bus Master and SDP Logic with the EPM5032 EPLD

This section describes a Micro Channel bus master interface with streaming data transfer capability. A data transfer rate of 80 Mbytes/sec can be achieved by using full 32-bit address and data buses in the interface design. The essential bus master and streaming data functions are implemented in an Altera EPM5032 MAX EPLD using thirteen inputs and seven outputs. The logic is implemented as a synchronous state machine timed by a 20-MHz clock. The EPM5032 works together with the Altera EPB2001 and EPB2002A EPLDs to provide a complete Micro Channel bus master interface. The features of this implementation are listed below.

- Bus control logic for basic and streaming data transfer
- Support for basic transfer cycles
- -Default (200-ns cycle)
- -Synchronous extended (300-ns cycle)
- -Asynchronous extended (>300-ns cycle)

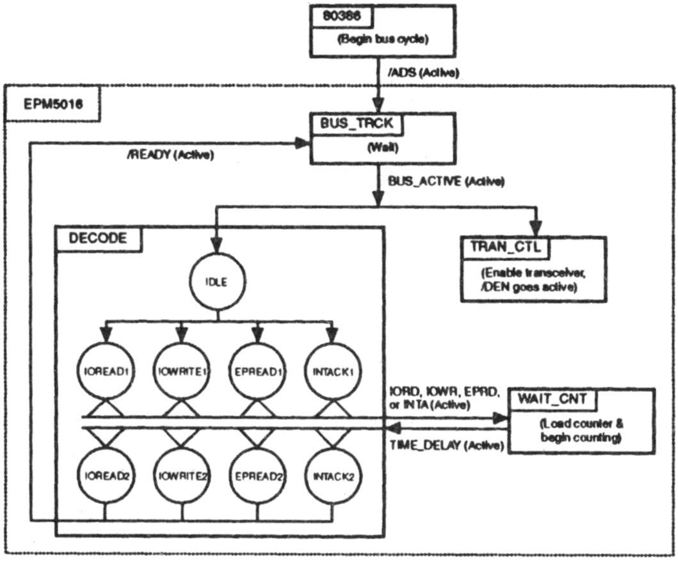

Figure 4.6.4.2 Bus Controller Flow Diagram

- Support for streaming data enable function (external POS bit)
- Support for streaming data cycles; (16-, 32-, and 64-bit)
- -Deferred streaming cycles
- -Data pacing
- -All modes of streaming data cycle termination
- Clock signal generation to track -SD STROBE during data pacing
- Bus master and SDP logic implementation as a synchronous state machine timed by a 20-MHz clock
- User-defined pin assignments allowed (except VCC and GND) and input polarities
- Available in 28-pin DIP, SOIC, or JLCC packages

Figure 4.6.5.1 illustrates the Micro Channel interface implemented with the EPB2001, EPB2002A, and EPM5032 EPLDs. The application-specific address control, address generation, and data steering functions are left to the designer for greater flexibility. The EPB2001 provides the I/O or memory-slave interface, and the EPB2002A provides DMA arbitration functions. The EPM5032 implements bus master control logic with SDP capability.

When the EPB2002A signals to the EPM5032 that the bus has been won, the EPM5032 generates the essential bus control signals. If streaming data capability is enabled (via any user-configurable POS register bit from an EPB2001 POS I/O pin)

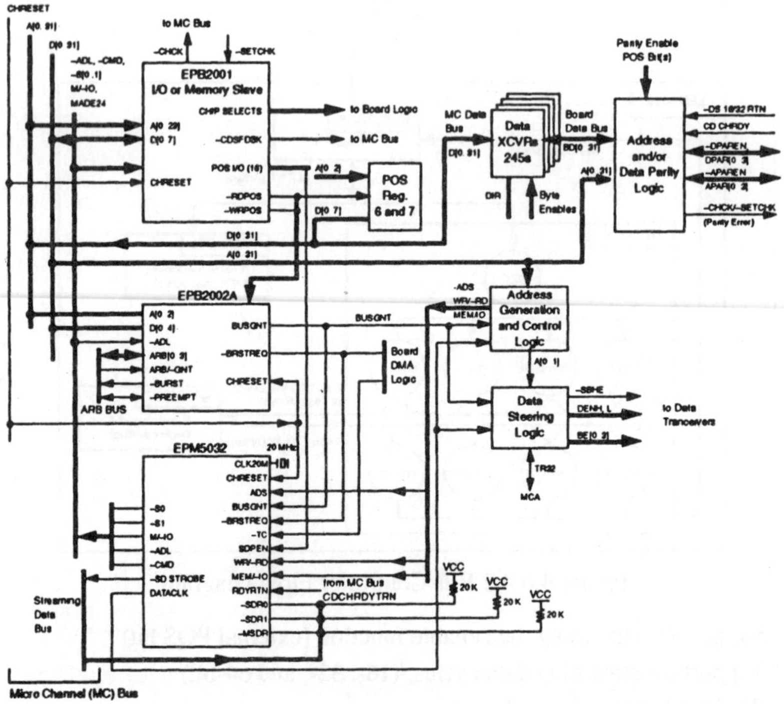

Figure 4.6.5.1 Micro Channel Bus Master Interface

the EPM5032 detects whether the slave can perform streaming operations and, if so, generates the -SD STROBE signal and DATA_CLK output. The DATA_CLK output tracks the -SD STROBE signal. It uses the RDYRTN signal to provide a "stretched out" clock for internal use on the board (e.g., for use by an address increment counter).

Bus control for burst transfers is indicated to the EPM5032 by the EPB2002A signals BUSGNT and -BRSTREQ. BUSGNT is an EPB2002A output which signals that the bus has been won. An active -BRSTREQ input to the EPB2002A indicates that the bus was requested by the adapter for burst transfers.

The -ADS input is required to determine when addresses are valid on the address bus. It is similar to the -ADS signal on the 80386 processor, which indicates the start of the cycle, and is used by the EPM5032 to latch the state of the WR/-RD and MEM/IO signals. After these signals are latched, the EPM5032 drives the -SO, -S1, M/-IO, -ADL, and -CMD signals. The -SO and -S1 status signals are decoded from the state of the WR/-RD signal; M/-IO is decoded from the state of the MEM/IO input.

The bus master and streaming data logic are implemented in AHDL as a synchronous state machine. Figure 4.6.5.2 shows a diagram of this state machine. It is timed by a

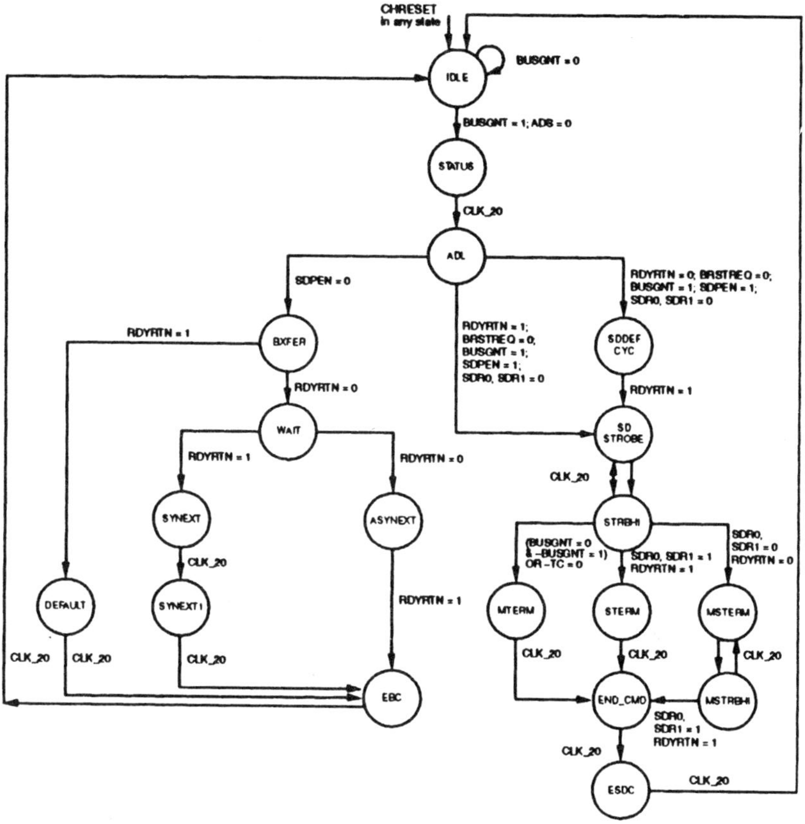

Figure 4.6.5.2 Bus Master SDP Controller State Machine

20-MHz clock fed on the CLK_20 input pin. Outputs of the state machine are decoded as functions of the present state. The next state is determined by the current state and inputs.

4.6.6 FIFO Controller Using an EPM7096

Interacting digital subsystems often consume and produce data at different rates and times. A First In/First Out (FIFO) buffer in a data path can bridge this rate and time mismatch so systems can function correctly. With the appropriate FIFO buffer, data written into the FIFO can be read in the order it was entered, but at an independent rate. This section describes how to use the 96-macrocell EPM7096 to design an 8 K x 16 bit FIFO.

Two types of FIFO buffers are available: shift FIFOs and pointer FIFOs. Shift FIFOs are typically constructed from a linear array of registers, where the number of registers is the same as the number of bits in the data word. Data to be stored in the FIFO is written to the first location in the register array. This data then "bubbles" to the last empty location nearest the output. When data has shifted to the last location of the register array, it can be read. The major drawback of shift FIFOs is fall-through time delay, which is the time required for valid data to propagate from the input to the output of an empty FIFO.

In contrast, pointer FIFOs do not have this fall-through time delay. Data is stored in a Random Access Memory (RAM) array. Read pointers store the next address to be read from this array; write pointers store the next address to be written to this array. If these pointers are made up of counters that roll over (i.e., restart at 0) at a value equal to the size of the FIFO, the FIFO structure becomes a circle. See Figure 4.6.6.1.

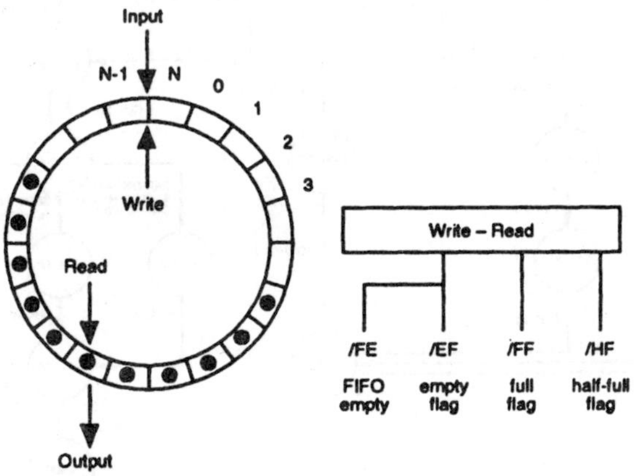

Figure 4.6.6.1 Circular Pointer FIFO

To enable the designer to determine whether the FIFO is full, empty, or at any point in between, the following Up/Down counter can be added to the design:

amount of data = (number of writes) - (number of reads)

Pointer FIFOs are available in a single package that contains read and write pointers and a RAM array. With this option, however, the size of the RAM arrays and the type of control functions available are limited. A designer can get around this limitation by using a standard static RAM device and implementing the control logic in an EPLD to maximize buffer-size options and meet custom interface-control requirements. The EPLD/RAM solution also significantly reduces cost. An 8 K x 16 bit FIFO implemented with RAMs and an EPLD costs less than half the price of a comparable off-the-shelf solution with two 8 K x 8 bit FIFOs.

Figure 4.6.6.2 shows a typical application of a pointer FIFO buffer implemented with

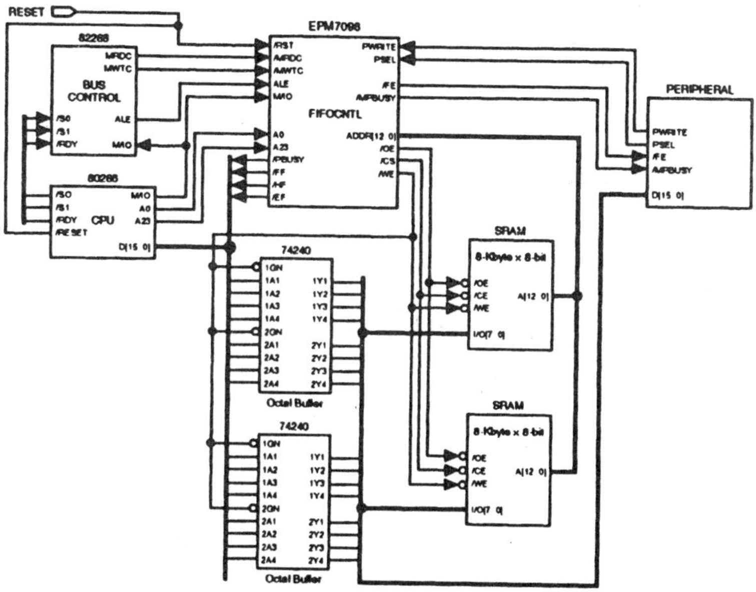

Figure 4.6.6.2 EPM7096 FIFO Application

an EPM7096 and two 8 K x 8 SRAMs. The input to the FIFO is a 10-MHz 80286 microprocessor. The FIFO controller interfaces to the /MRDC, /MWTC, M/IO, and ALE signals from the 80286 and 82288 bus controllers. Chip select is generated by decoding the address lines A0 and A23.

The microprocessor can read the FIFO status from the four active-low tri-stated lines that are connected directly to the microprocessor's data bus: /EF (empty flag), /HF (half full), /FF (full flag), and /PBUSY (peripheral busy). The microprocessor bus and the peripheral bus are isolated by a pair of 74240 octal buffers controlled by the /WE output of the EPM7096 FIFO controller.

In this example, an arbitrary peripheral is connected to the output of the FIFO. This peripheral selects the FIFO with the PSEL (peripheral select) line and strobes data out of the FIFO with the PWRITE (peripheral write) line. The peripheral determines the status of the FIFO by reading /FE (FIFO empty) and /MPBUSY (microprocessor busy) signals.

The two 8 K x 8 bit static RAM devices store up to 8192 16-bit data words (the same number of bits as the 80286 microprocessor's data bus). The EPM7096 FIFO controller provides the RAM with all required control and address lines, including /OE (output enable), /CS (chip select), /WE (write enable), and address lines A0 to A12.

Figure 4.6.6.3 shows the block diagram for the EPM7096 FIFO controller. Two 13-bit

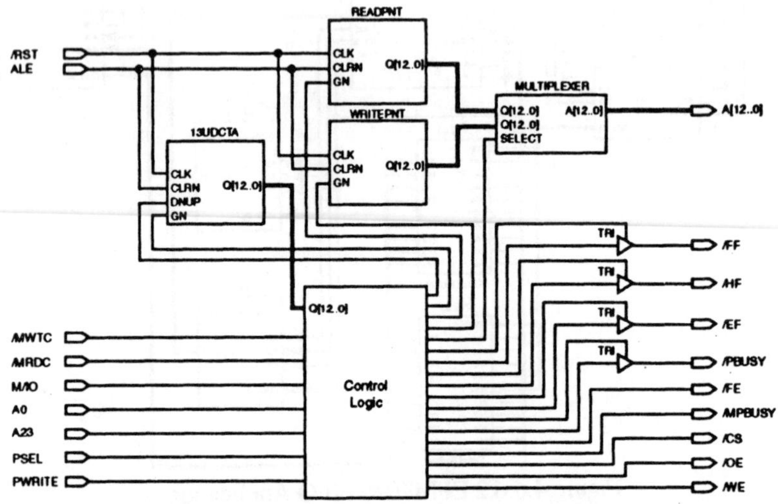

Figure 4.6.6.3 EPM7096 FIFO Controller

counters provide the read and write pointers for the EPM7096. The read pointer counter contains the address where the next read data is stored; the write pointer counter contains the address where the next write data is stored. A third 13-bit counter, 13UDCTA, determines FIFO status (13 bits define 8192 unique addresses).

When a FIFO read or write operation is required, a 26- to 13-bit address-pointer multiplexer selects the read or write pointer outputs and presents them to the RAM address bus. The control logic block decodes FIFO requests, provides FIFO status outputs, and controls the address-pointer multiplexer.

4.6.7 Integrating an Intelligent I/O Subsystem with a Single EPM5130 EPLD

When higher system performance is necessary, many designers first consider a faster microprocessor or a new microprocessor architecture. In many cases, however, sufficient speed can be achieved through an intelligent I/O subsystem that has been optimized for a particular task. Transferring I/O processing to intelligent subsystems also allows the system processor to dedicate more processing power to primary system functions.

This section describes how the custom logic requirements of such a subsystem can be integrated into a single Altera EPM5130 MAX EPLD, replacing over 75 standard TTL packages.

Figure 4.6.7.1 shows a subsystem using the T1 serial coprocessor implemented in an EPM5130 EPLD. The serial coprocessor consists of two T1 serial transmitters, control and address generation logic for a buffer RAM, and the I/O subsystem control logic. The system processor writes data for serial transmission through the T1 serial coprocessor into the buffer RAM. When data transfer is complete, the system processor signals to the serial coprocessor that data can be transmitted. The serial coprocessor then transfers the data to the transmitters for serialization. The buffer RAM control logic also features error detection and correction. Data can be sent over either T1 serial channel with individual variations in protocol. Once all of the data in buffer RAM is transferred, the processor is interrupted, and the cycle can repeat.

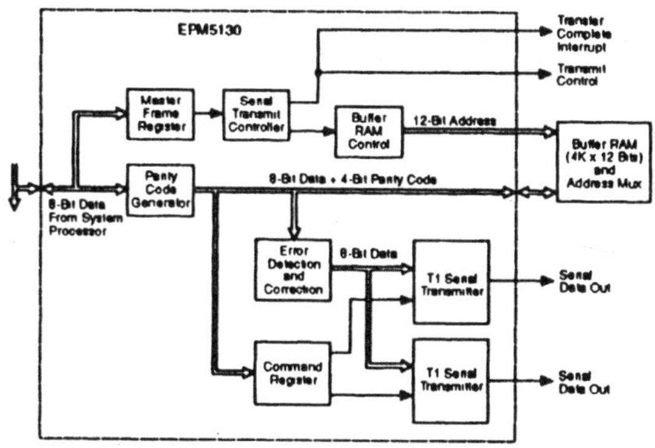

Figure 4.6.7.1 T1 Serial Coprocessor Block Diagram

The T1 serial coprocessor is a useful example of an intelligent I/O subsystem design because it contains the types of logic, such as decode and control state machines, that are common to most digital designs. The error detection and correction, RAM control and address generation, and parallel-to-serial conversion features of the T1 coprocessor are also widely used in system design. Furthermore, digital data communication of both voice and data plays an increasingly important role in modern systems.

Although this design example shows a T1 serial transmission application operating at 1.544 MHz, the same concept can be applied to create high-performance subsystems for applications such as local area networks or disk controllers. The EPM5130 EPLD supports serial data rates of up to 50 megabits per second.

4.6.8 Controlling Complex CCD Imaging Systems with the EPS464 EPLD

Advances in VLSI-scale microelectronics have allowed manufacturers of video and optical equipment to replace optical film storage media with arrays of light-sensitive electronic imaging elements. These video systems use analog charge-coupled device (CCD) technology to retain image data, as well as a high-speed digital control system to convert the image into a standard video format.

The Altera EPS464 Synchronous Timing Generator (STG) is an ideal device for implementing the necessary digital control logic. The STG is well suited for generating complex waveforms, control signals, and state machines. The EPS464 EPLD contains 64 macrocells optimized for timing and waveform synthesis applications. Thirty-two of these macrocells are connected to I/O pins, while the other 32 are available for buried logic such as state machine registers and user-defined n-bit counters. A global bus feeds all macrocells within the device allowing system speeds of 66 MHz.

The complex digital control system for a hand-held video camera requires a combination of high speed and high density. Not only must the system retrieve image data from the imaging element, it must also process the analog data into a standard format for broadcast or storage on magnetic tape. All of the functional blocks must work together and be perfectly synchronized.

Most hand-held cameras use a CCD element as the image capture medium. The CCD array does an excellent job of digitizing the incident images at high speed; however, the larger the resolution of the array, the faster the digital control system must operate. An array with a resolution of 910 x 525 pixels generates over 450,000 data points for every image. To meet the NTSC standard of 30 images per second, the control system must operate at 14.318 MHz. The combination of high density, high performance and architectural flexibility of the EPS464 is an ideal solution to this problem.

Figure 4.6.8.1 shows a typical implementation of a controller for a hand-held 8-mm video camera. The application uses a 910 x 525-bit CCD imaging array and circuitry to format the image data into an NTSC video standard.

The design contains three main blocks: a horizontal (X) counter, a vertical (Y) counter, and a synchronous signal generator (SSG). In Figure 4.6.8.1 the two counters are shown as timing generators which are used to format the video display into pixels per line and lines per screen. The SSG uses these counter values to generate the necessary control patterns and formatting signals.

The size of the counters varies, according to the desired video format and the resolution of the display. The NTSC format specifies an interlaced display scheme; therefore, the system must count to one-half of the 525 lines in the display, or 262.5 lines. However, the count is difficult to control since it depends on half clock cycles. To solve this problem, the screen is divided into a series of 1050 half-lines, with each frame using 525. By offsetting the display by half a horizontal line and displaying the two "half-pictures" in rapid succession, the interlace appears fluid, and the result is a very good representation of the original picture. This approach takes advantage of the physiology of human vision: although the human eye is good at distinguishing a wide range of colors, it is not so good at distinguishing small or rapid movements. Consequently, an image updated 30 times per second appears to have fluid motion.

Given these system requirements, the horizontal counter needs 10 bits to count the

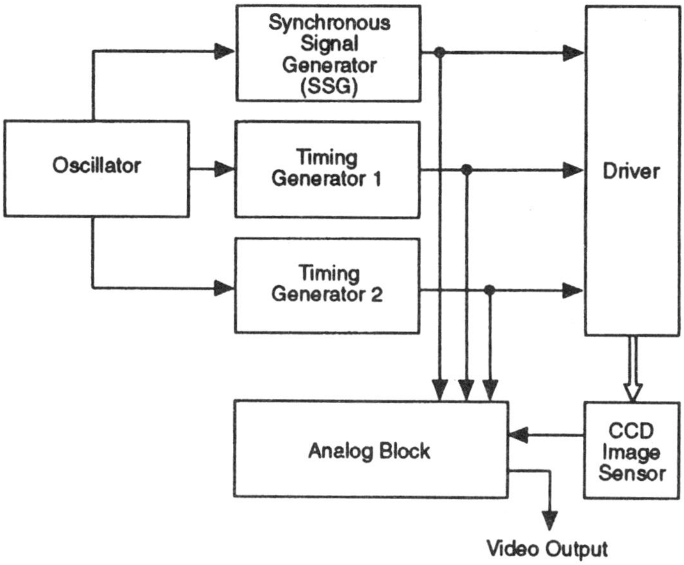

Figure 4.6.8.1 Typical CCD Controller

910 pixels per line and the vertical counter needs 11 bits to count the 1050 lines per screen. By using the 1050 half-line approach, the counters will use integers rather than a cumbersome scheme of half clock cycles.

The NTSC standard has three key specifications: lines per screen, line period, and screens per second. Since each image must be synchronized by the display hardware, and any variation from the standard would cause the image to roll on the screen, the lines-per-screen specification is fixed at 525. However, not all 525 lines are actually shown on the display. Twenty are used for formatting and equalization intervals, while the remaining 505 are dedicated to image data.

The second specification is the amount of time that is permitted to display a line. Each line is encoded during a 63.556s period established by the incoming video signal. To encode the 910 horizontal pixels in the application, the control system must divide each horizontal time period into 910 individual segments and sample the analog video signal at each of those points. Each sample corresponds to the clocking out of the next word of analog data from the CCD array. With a horizontal time period of 63.556s, the sampling rate is 14.318 MHz.

The third NTSC specification requires that an image must be updated 30 times per second, requiring a refresh every 33.33 ms. Since the NTSC is an interlaced format with two frames per image, the actual frame refresh rate is 60 Hz. This rate is easily achieved with the EPS464 EPLD as the digital control system.

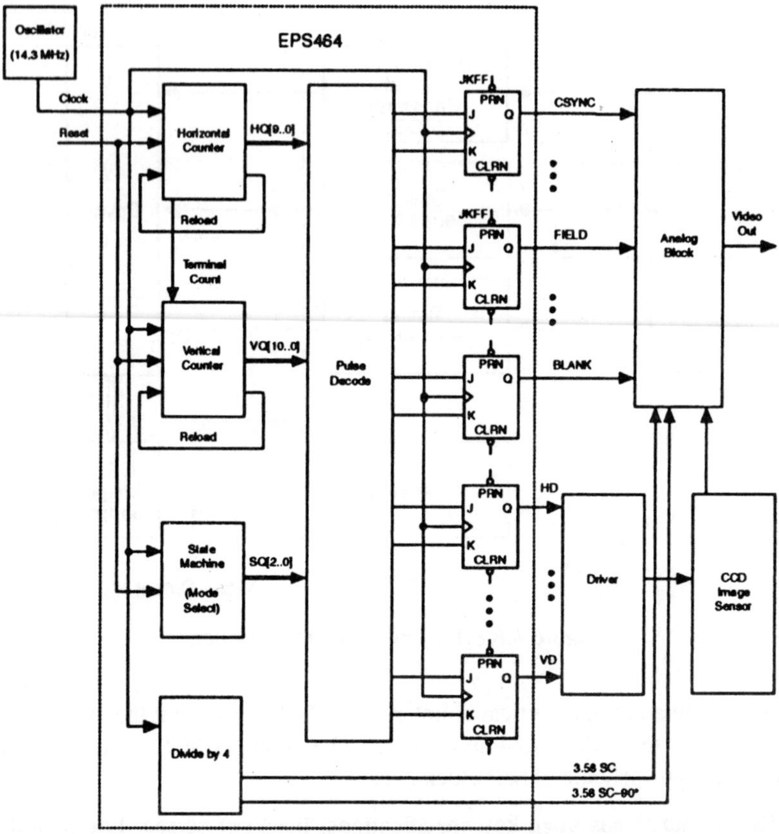

Figure 4.6.8.2 Top Level of the CCD Imaging System Control Application

The final video output is an analog signal that combines analog voltages representing intensity and color information with digital formatting pulses. Although digital and analog levels are combined with a special-purpose analog mixing device, mixing is actually performed under the control of the digital system. The result is a single composite video signal for broadcast or storage on tape.

Figure 4.6.8.2 shows this application implemented in the Altera EPS464 EPLD. All five subfunctions (horizontal counter, vertical counter, state machine, pulse decode module, and clock divider) fit inside a single EPS464 and can be easily modified to meet the requirements of any application.

4.7. References

Ahanin, B. and Lytle, C., "Methods and Apparatus for Facilitating Scan Testing of Asynchronous Logic Circuitry," U.S. Patent Application (pending).

Altera Application Brief AB 76, "Using Expanders to Build Registered Logic in EPM5000-Series MAX EPLDs," October 1990, Version 2.

Altera Application Note AN 22, "Designing with AHDL," Oct. 1990, Version 1.

Altera, "Altera EDIF Writer Specification Version 3.0," Specification Number 16C-01199, Dec. 1991.

Altera, "Valid/Cadence Design Kit for PC Users," Altera Software Utility.

Altera, "Viewlogic Design Kit for PC Users," Altera Software Utility.

ANSI/EIA - 548-1988, "Electronic Design Interchange Format Version 2 0 0."

Birkner, J. et al. (editor), *Programmable Array Logic Handbook, First Edition*, Monolithic Memories Inc., 1978.

Datta, S. and Kung, E., "EDIF: Its Contribution to the PLD Design Flow," *The PLD Design Conference and Exhibit Proceedings*, 1991.

Fiduccia, C.M. and Mattheyses, R.M., "A Linear-Time Heuristic for Improving Network Partitions," *19th Design Automation Conference*, Paper 13.1, pp 175-181.

Hartmann, R., "Estimating Gate Complexity of Programmable Logic Devices," *VLSI Design Magazine*, pp 100-102, May 1984.

Kernighan, B.W. and Lin, S., "An Efficient Heuristic Procedure for Partitioning Graphs," *The Bell System Technical Journal*, Feb. 1970, pp 291-307.

Kitson, B. et al. (editor), *Programmable Logic Array Handbook*, Advanced Micro Devices Inc., 1984, pp. 2-29 through 2-40.

Munoz, R. R. et al., "Automatic Partitioning of Programmable Logic Devices," *VLSI Systems Design*, Oct. 1987, pp. 74-76, 78 and 86.

Wolf, Stanley, *Silicon Processing for the VLSI Era*, Lattice Press, CA, 1990, pp 623-635.

Wong et al., "Programmable Logic Device with Array Blocks Connected Via Programmable Interconnect," U.S. Patent 4,871,930, Oct. 3, 1989, Col 2, line 43 et seq.

Index

A

ABEL 143
Act1 110–113
Act2 113–117
Act3 117–118
ACTMAP 140
ADL 143
Ahanin 202
Ahrens 97, 118
Allen 97, 99
Altera 191, 216, 235
amorphous silicon 100
analog FPGA 4
ANSI 205
antifuse 99
    dielectric 100
    F antifuse 109
    ONO 100
    resistance 100
Apple 82
array-based architecture 184
automatic test vector generation 202

B

binary counter 71
binary decision diagram 129
Birkner 100, 171
bit-per-state 160
    see also one-hot 69
bit-slice 49
Blkmake 57, 59
Boolean equations 128
boundary scan 45, 80
Brayton 129
Breuer 58
Britton 15
buffering 25

C

capacity 51, 124
carry chain 44, 219
carry hard macro 156
carry logic 44, 70
carry-propagation 153
carry-select 154, 160
Carter 15, 97, 99
cascade 26
cascade chain 219
channel density 104
Chen 101
Chene 57
Chiang 101, 102, 103
Chortle 55, 57
Chow 3, 15
cip, see pip 20
CLB 19, 22, 29, 36, 44
    see also logic module 113
    see also macrocell 180
CLBMAP 64

Cliff 15
clock routing 33, 49
C-module 113, 117
CMOS 18
Compiler 208
Complex PLD 3
configurable display device 82
configurable interconnect point, see pip 20
configurable logic block, see CLB 19
configurable memory 45
configuration
    see programming 52
configuration bitstream 22
configuration memory 15
Cong 55
controllability 52
counter 70, 144, 226
Cox 75
CPLD 3, 171
crossbar switch 22

D

Datta 205
debugging 84, 131
density 10, 23
design migration 11, 200
digital signal processing 77, 164
direct addressing 120
direct connect 26, 71
    see also cascade 219
direct connection 32
distributed arithmetic 77
DMA controller 85
Donath 29
double-entry symmetry 116
double-length lines 46
dual feedback 182, 185
Dunlop 58
duplicated logic 57, 69

E

Ebeling 3, 15
EDIF 211
ElAyat 97
electronic scaffolding 84
ElGamal 29, 97, 106, 107, 109, 111, 118
EP1810 184–187
EPROM 173
expander 189, 190, 198, 229
expander product term 189

F

fan-in 134
fan-out 133
FastTrack 219, 222
Fiduccia 58, 216
Field Programmable Gate Array, see FPGA
Field-Programmable Analog Array (FPAA) 4
fine-grain vs. coarse grain 27
fitting 172, 204, 216
    see also partitioning 54
flexible processor 90
flip-flop 28, 69
floppy disk 82
force-directed relaxation 60
FPGA
    advantages 4
    architecture 3
    cost comparison with MPGA 9
    definition 2
    density 8
    design migration to MPGA 11
    disadvantages 8
    speed 9, 10
    taxonomy 3
Francis 55, 57
Frankle 62
function generator, see lookup table 20

Furtek 15

G
Ganglion 75
gate array 1
general-purpose interconnect 46
Gerzberg 97, 100
global clock 70
global interconnect 33, 48
globally connected 184
gobbler 138
Goto 60
Graham 100, 118
graphic editor 208
Greene 26, 107
grow back 99

H
Hamdy 97, 100
hard macro 137
Hartmann 171
Hartoog 58
Hashimoto 104
Hastie 15
Hauck 15
Heller 29, 49
high-performance counter 150
Hill 15, 29
Holley 67
Holmberg 100
Hsieh 15, 43, 97

I
I/O block 182
if-then-else DAG 129
in-circuit verification 6, 54
incremental placement 67, 130
incremental routing 67
input segment 108
IOB 30, 37, 45
    see I/O block 22
iterative design 70

J
JEDEC 216
jitter-bounded PLL 165
JTAG, see boundary scan 45

K
Karplus 55, 129
Kawana 15
Kean 15
Kernighan 216
Kirkpatrick 60
Kitson 190

L
LAB 188, 197, 221
    see also CLB 22
    see also logic module 113
leakage current 100
Lee 4
left edge algorithm 104
levels of logic 133
library 54, 66
Library Mapping Files 211
life cycle 7
Lim 100
limit-bumping algorithm 62
Lisa 4
loadable counter 144
logic array 180
logic array block 188
    see LAB 187
logic module 113
    see also CLB 22
    see also macrocell 180
logic optimization 66, 213
logic synthesis 63, 172, 204
Logic Synthesizer 213
long line 32
lookup table 20, 69, 219

Lorenzetti 104
LPM 67
LSI 130
LUT 20
Lyons 17

M

macro, combinable 159
macrocell 180, 186, 189, 197
    see also CLB 22
macrofunction 179
macrofunction library 211
magic box
    see switchbox 22
Mailhot 129
mapping 129
    see fitting 216
    see partitioning 54
Marr 106
Mask Programmed Gate Array (MPGA) 1
Mask-Programmed Logic Device 200
mask-programmed logic device (MPLD) 200
Maunder 80
MAX5000 187–195
MAX7000 195–200
memory 45
memory cell in SRAM FPGA 15
message processor 207
Michael 0
microparameter 224
min-cut 58, 216
minimization 204
Mintzer 77
MIS-II 55
MIS-pga 55
module utilization 124
MPGA 1, 29
multi-chip partitioning 206
multi-gate macros 134
multiplexer 21
multiplication 75, 77
Munoz 190
Murgai 55, 129
Muroga 15

N

nearest-neighbor connections
    see direct connect 26
neural network 75
Nilsson 61
NRE 5, 7

O

observability 52
one-hot 69
    see also bit-per-state 160
output enable 142
output pin swapping 34
output segment 109

P

pad limited 10
PAL 184
PALASM 140, 143
Pamela 2
parallel logic expander 198
partitioner 216
partitioning 54, 216
    see also fitting 216
pass transistor 21
Penfield 62
performance 127
Perle-0 88
Perle-1 88
PIA 189, 193
pin assignment 130
pin swapping 34, 61
pip 20, 25, 26
pipelining 70
placement 41, 59, 67, 130
    directional bias 60
    incremental 68

PLD 29
PLICE 100
position tracker 84
power consumption 19
Preas 109
pre-scaled counter 150
probe points 65
product term 177
program 22
programmability
    effect on density 8
    effect on speed 9
Programmable Array Logic (PAL) 1
programmable interconnect array 189, 192, 199
programmable interconnect point, see pip 20
Programmable Logic Device (PLD) 1
Programmable Multi-level Devices (PMD) 3
Programmer Object Files 216
programming 52, 118, 131, 173
    overhead 8, 10
    re-programming 52
    time 131
programming yield 6, 18, 173, 183
prototype 67, 84, 87
pulse-steal PLL 164

Q

Quickturn Systems 87

R

RAM 45
readback 52, 90
reconfiguration 52
    see reprogramming 52, 79
Report File 212
repowering buffers 32, 40, 49
reprogramming 17, 70, 76, 79–83
resource limits 29
ripple-carry adder 71
Roesner 100
Rose 27, 28
Rosendahl 80
routability 25, 29
routing 61, 67, 104, 130
    global route 61
    maze route 61
Rubenstein 62
Rudell 129

S

Sawkar 55
scan-testing 201
schematic editor 53
Schlageter 97
Sechen 59, 60
security 53, 182
segment 22, 32
segmented routing channel 26, 106
series pass transistor addressing 120
74AS161 138
shared logic expander 197, 198
simulated annealing 60
simulation 67
Simulator Netlist File 201
slack 62
sloppy design 9
Smith 97, 99
S-module 113, 117
    combinable 117
sneak path 123
soft macro 135, 137
Soukup 61
speed 24, 43, 51
speed binning 7
Splash 89
state-machine generator 53
Steiner tree 109
Stopper 100
sub-families 49
sum-of-products 177

switchbox 22
Symbol Editor 208
synchronous binary counter 144
synthesis 53, 67, 141
    technology specific 63

T

TA161 138
table lookup
    see lookup table 20
Tan-Nguyen 82
tape drive 82
technology mapping
    see partitioning 54
telecommunications 164
testing 6, 18, 79, 118
three-state 141
three-state buffers 39
    placement constraint 59
    placement constraints 60
threshold drop 21
time-dependent dielectric breakdown 102
timing-driven placement 130
TLU
    see lookup table 20
tooling 4
Trimberger 43, 57
two-segment routing 107

V

video controller 83
virtual hardware 90
volatility 17

W

Walters 165
waveform editor 207, 209
Whitney 97
Whitten 97, 100
wiring channel 22
Wolf 173
Wong 97, 106, 184
Woo 55

X

XACT Design Editor 64
X-BLOX 63
XC2000 29–35, 40
XC3000 35–43
XC4000 43–51
XC4000A 49
Xilinx 17
Xnfmap 56